CONSCIOUSNESS AND ROBOT SENTIENCE

Series on Machine Consciousness

ISSN: 2010-3158

Series Editor: Antonio Chella *(University of Palermo, Italy)*

Series on Machine Consciousness – Vol. 2

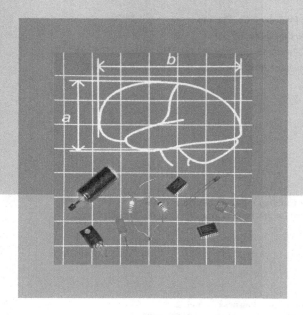

CONSCIOUSNESS AND ROBOT SENTIENCE

Pentti O Haikonen

University of Illinois at Springfield, USA

World Scientific

NEW JERSEY · LONDON · SINGAPORE · BEIJING · SHANGHAI · HONG KONG · TAIPEI · CHENNAI

Published by

World Scientific Publishing Co. Pte. Ltd.

5 Toh Tuck Link, Singapore 596224

USA office: 27 Warren Street, Suite 401-402, Hackensack, NJ 07601

UK office: 57 Shelton Street, Covent Garden, London WC2H 9HE

British Library Cataloguing-in-Publication Data
A catalogue record for this book is available from the British Library.

Series on Machine Consciousness — Vol. 2
CONSCIOUSNESS AND ROBOT SENTIENCE

Copyright © 2012 by World Scientific Publishing Co. Pte. Ltd.

ISBN 978-981-4407-15-1

Printed in Singapore by World Scientific Printers.

Dedication

This book is humbly dedicated to you, my respected reader and to the international community of machine consciousness researchers.

Preface

Many excellent papers on machine consciousness were presented at the AAAI Fall Symposium on AI and Consciousness at Arlington in 2007. Therefore, at the concluding moments of one presentation the harsh comments of one philosopher were a disturbing shock. The comments were presented in a frank way, but the most disturbing thing was that the philosopher was right. Stevan Harnad lamented that none of the presented papers really addressed the one and only real issue of consciousness and therefore did not deliver the promised goods. This statement was not fully justified, because there were papers that treated the issue, even though not necessarily in a sufficient or even a correct way. Harnad noted that there are many proposed ways to build a machine or a write a computer program that is claimed to be conscious. But alas, none of these approaches really solve the real problem of consciousness and therefore any claims of consciousness in these machines and programs are totally unfounded. The author has no other option than to agree with Harnad.

The real problem of consciousness is recognized by many philosophers in one form or another and many designers of potentially sentient robots are at least vaguely aware of it. But, the real problem has appeared so hard that it has been easier just to ignore it and hope that it will be automatically solved, as soon as a sufficient system complexity is achieved. It will not.

This book is different from most of the contemporary books on machine consciousness. The real problem of consciousness is taken as the starting point and it is explained, not explained away. The rest is developed from the conclusions from this investigation. This book is the third one in the author's trilogy of machine consciousness books, the previous ones being "The Cognitive Approach to Conscious Machines", Imprint Academic UK 2003 and "Robot Brains; Circuits and Systems for Conscious Machines", Wiley UK 2007. A reader will notice that similar themes are presented in these books. These books augment each other; the first book presents background contemplations from philosophy and cognitive sciences and the second book presents material for engineers. This book amplifies, clarifies and explains the real and fundamental issues and practical aspects of machine consciousness and includes a presentation of the author's experimental cognitive robot XCR-1.[*]

I would like to thank Prof. Peter Boltuc, Prof. Antonio Chella and Mr. Dylan Drummond for their expert comments and valuable suggestions about the text.

Finally, I want to thank my wonderful, ever young wife Sinikka for her support and encouragement; I can move mountains, if you only hold my hand. Special thanks go also to my media artist son Pete, whose metaphysically captivating and uplifting techno-instrumental composition *Untitle 'Em* with its strong spectrum of amodal qualia gave me inspiration and strength to continue during the longest hours of this effort.

Pentti O A Haikonen

---

[*] Demo videos of the robot XCR-1 can be seen at
http://www.youtube.com/user/PenHaiko/videos

Contents

Chapter 1

Introduction

1.1. Towards Conscious Robots

Why is it so hard to make computers to understand anything and why is it equally hard to design sentient robots with the ability to behave sensibly in everyday situations and environments? The traditional approach of Artificial Intelligence (AI) has not been able to solve these problems. The modest computing power of the early computers in the fifties was seen as the main limiting factor for Artificial Intelligence, but today, when hundreds of gigabytes of memory can be packed into miniscule microchips, this excuse is no longer valid.

Machines still do not understand anything, because they do not operate with meanings. Understanding everyday meanings calls for embodiment. The machine must be able to interact with its environment and learn how things are. We humans do this effortlessly, we see and understand our environment directly and can readily interact with it according to the varying requirements of each moment. We can do this, because we are conscious. Our existing robots cannot do this.

Traditional Artificial Intelligence has not been able to create conscious robots (except in science fiction movies) and it is very likely that it never will (for the reasons explained later on in this book), therefore novel approaches must be found. We know, how conscious subjects behave and we could make machines imitate conscious behavior using the traditional AI methods. However, the mere imitation of conscious behavior is not enough, imitation has its limits and may fail any moment. The only universal solution would be the creation of truly conscious machines, if we only knew what it takes to be conscious.

1

The philosophy of mind has tried to solve the mystery of consciousness, but with limited success. Contradicting opinions among philosophers abound, as was demonstrated by Aaron Sloman's target paper [Sloman 2010] and the responses to it in the International Journal of Machine Consciousness Vol. 2, No 1.

Psychology has not done much better. Neurosciences may be able to associate some neural activities with conscious states, but how these neural activities could give rise to the experience of being conscious has remained unexplained.

Could engineering do better? Engineers are familiar with complex systems. They know, how components can be combined into systems that execute complicated functions. The creation of a cognitive machine should thus be just another engineering project; feasible and executable as soon as the desired functions are defined. These definitions could be delivered by cognitive sciences. No doubt, machines that behave as if they were more or less conscious can be eventually created. But, were these machines really conscious? Would they have the experience of being conscious? If not, then no real consciousness has been created and the machines would only be mere replicas of the original thing, looking real, but not delivering the bullet.

The bullet must be delivered and conscious robots must be aware of their own thoughts and their own existence and know what they are doing. Unfortunately, there has been no engineering definition for the experience of being conscious. Philosophers have been pondering this question for couple of thousand years and have come to the conclusion that the phenomenon of consciousness seems to involve a problem that is hard or even impossible to solve in terms of physical sciences. This issue is known as the mind-body problem. The mind involves the experience of being conscious; how could this be explained with the application of the laws of physics? The natural laws of physics are able to explain the workings of energy and matter, electromagnetic fields, atoms and electrons and eventually the whole universe, for that matter. Yet the natural laws of physics have not been able to explain consciousness. Therefore, is consciousness something beyond energy and matter?

The central hypothesis beyond this book proposes that consciousness is neither energy nor matter and therefore it cannot be explained by the physical laws about energy and matter. However, consciousness is achievable by the application of energy and matter, because we already have an example of this kind of a conscious system; the human mind. Thus, the experience of being conscious seems to be a property of certain kinds of perceptual systems. Therefore, the explanation for consciousness would be found at the system level and the creation of robots with conscious minds should be possible. However, before the construction of conscious machines the real problem of consciousness must be identified, isolated and analyzed.

1.2. The Structure of This Book

This book begins with the review of the fundamental philosophical issues of consciousness. This treatment leads to the identification of the one and only real problem of consciousness. It is shown that this problem is related to perception and qualia, which are then discussed in detail. The existence of amodal qualia is proposed based on the known concept of amodal features and it is proposed that amodal qualia can give some insights into the phenomenal nature of qualia.

The relation of emotions and inner speech to consciousness is discussed next.

Can machines have qualia? It is argued that qualia are a mandatory prerequisite for human-like machine consciousness. Systems without qualia are not truly conscious. Next, some preconditions for machine qualia are proposed.

How do we know that a person or a machine is conscious? Some proposed tests exist and are presented and discussed here.

The identified preconditions for conscious cognition lead to the requirement of a perceptive system that combines sub-symbolic and symbolic information processing. Associative information processing with associative neural networks and distributed signal representations is introduced as a method for sub-symbolic processing that inherently facilitates the natural transition from sub-symbolic to symbolic processing.

Conscious robot cognition calls for information integration and sensorimotor integration and these lead to the requirement of an architecture, the assembly of cross-connected perception/response and motor modules. The Haikonen Cognitive Architecture (HCA) is presented as an example of a system that would satisfy the identified requirements.

Modern brain imaging technology seems to allow at least limited mind reading. It is proposed that the HCA might be used to augment mind reading technology. One already implemented example is cited.

Many cognitive architectures have been proposed lately and the comparison of their different approaches with the HCA would be interesting and in this way the approach of this book could be put in a wider perspective. However, a complete comparison is not feasible here, therefore a smaller review is attempted and the compared cognitive architectures are the Baars Global Workspace architecture and the Shanahan Architecture, as these are well-known and share many representative features with several other architectures.

Finally, as an example of a practical implementation of the Haikonen Cognitive Architecture, the author's experimental cognitive robot XCR-1 is presented.

At the end of each chapter, where feasible, a chapter summary is provided for easy assimilation of the text. The concluding chapter of this book summarises the explanation of consciousness, as proposed by the author. This summary should be useful and should reveal the main points quickly.

Chapter 2

The Problem of Consciousness

2.1. Mind and Consciousness

What is conscious mind? This question has been studied by philosophers since old ages and no clear and definite answers have been found. The popular answer is of course that mind is what we have inside our head; our observations, thoughts, imaginations, reasoning, will, emotions and the unconscious processes that may lie behind these. Our mind is what makes us a person. The strange thing about the mind is that it is aware of itself; it is conscious.

It is useful to note that "mind" is an invented collective concept which encompasses many aspects of cognition and as such does not necessarily refer to any real single entity or substance, which may not really exist. However, in the following the word "mind" is used in the popular sense; mind is the system of perceptions, thoughts, feelings and memories, etc. with the sense of self.

There is concrete proof that the brain is the site of the mind. However, when we talk about mental processes, the processes of the mind in terms of cognitive psychology, we do not have to describe the actual neural workings of the brain (in fact, at the current state of the art, we are not even able to describe those in sufficient detail). Thus, the separation between the mind and the brain would seem be similar to the separation between computer software and hardware. However, the mind is not a computer program. It is not a string of executable commands that constitute an algorithm. It is not even an execution of any such program. Nevertheless, the apparent possibility to describe mental processes without resorting to the descriptions of neural processes would suggest that mental processes could be executed by different machineries, just like a computer program can be run in different hardware, be it based on relays, tubes, transistors or microcircuits. This leads to the founding hypothesis of machine consciousness; conscious agents do not have to be based on biology.

It is possible that we will properly understand mind only after the time when we will be able to create artificial minds that are comparable to their human counterpart. But, unfortunately, in order to be able to create artificial minds we would have to know what a mind is. This is a challenge to a design engineer, who is accustomed to design exotic gadgets as long as definite specifications are given. In this case the specifications may be fuzzy and incomplete and must be augmented during the course of construction.

Real minds are conscious, while computers are not, even though the supporting material machinery should not matter. Therefore, the mere reproduction of cognitive processes may not suffice for the production of conscious artificial minds. Thus, there would seem to be some additional factors that allow the apparently immaterial conscious mind to arise within the brain. These factors are related to the real problem of consciousness.

2.2. The Apparent Immateriality of the Mind

Is the mind some kind of immaterial process in the brain? When we work, we can see and feel our muscles working and eventually we get tired. There is no doubt that physical work is a material process. However, when we think and are aware of our environment, we do not see or feel any physical processes taking place. We are not aware of the neurons and synapses in the brain, nor the electrical pulses that they generate. We cannot perceive the material constitution and workings of the brain by our mental means. What we can perceive is the inner subjective experience in the forms of sensory percepts and our thoughts. These appear to be without any material basis; therefore the inner subjective experience would seem to be immaterial in that sense. Indeed, our thoughts and percepts about real external objects are a part of our mental content and as such are not real physical objects. There are no real people, cars and streets or any objects inside our heads, not even very, very small ones, no matter how vividly we imagine them; these are only ideas in our mind. Ideas are not the real material things that they represent, they are something else. Obviously there is a difference between the real physical world and the mind's ideas of the same. The real physical world is a material one, but how about the world inside our head, our percepts, imaginations and feelings, the mental world? The philosophy of the mind has tried to solve this problem since old ages, but the proposed theories and explanations have led to further problems. These theories revolve around the basic problems, but they fail, because their angle of approach is not a correct one. In the following the problems of the old theories are highlighted and a new formulation of the problem of consciousness is outlined.

2.3. Cartesian Dualism

As an explanation to the problem of the conscious mind, The Greek philosopher Plato argued that two different worlds exist; the body is from the material world and the soul is from the world of ideas. Later on René Descartes suggested in his book *The Discourse on the Method* (1637) along the same lines that the body and the mind are of different substances. The body is a kind of a material machine that follows the laws of physics. The mind, on the other hand, is an immaterial entity that does not have to follow the laws of physics, but is nevertheless connected to the brain. According to Descartes, the immaterial mind and the material body interact; the immaterial mind controls the material body, but the body can also influence the mind. This view is known as *Cartesian Dualism*. Cartesian Dualism is also *Substance Dualism* as it maintains that two different substances, the material and the immaterial ones exist.

Is the dualist interpretation correct and do we really have an immaterial mind? This is a crucial issue for machine consciousness. Conscious machines should also have a mind with the presence of apparently immaterial subjective experience more or less similar to ours.

The artificial mind should thus contain mental content, which would appear to the machine as immaterial, without the perception of any underlying material processes. Machines are material; therefore, if real minds were immaterial, then the creation of immaterial machine minds with the material means that are known to engineers today, would be impossible. The dualist view of mind is illustrated in Fig. 2.1.

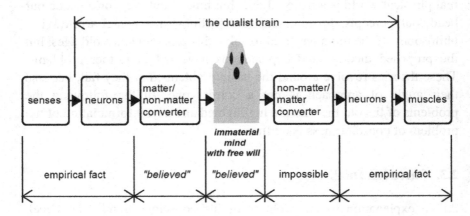

Fig. 2.1. The dualist view of mind. Senses provide information about the material world in the form of material neural signals. These affect the immaterial mind, which, in turn controls the material neurons that operate the muscles. In this process a non-matter/matter converter is needed. This converter would violate physics as it would have to produce matter and/or energy out of nothing.

The dualistic approach may at first sight seem to succeed in explaining the subjective experience of the immaterial mind with the proposed existence of two different substances; the material substance of the physical world and the immaterial substance of the mind. Unfortunately, while doing so it leads to the *mind-body interaction problem*; how could an immaterial mind control a material body?

It is known that the brain receives information about the material world via sensors; eyes, ears, and other senses that are known to be material transducers. The information from these sensors is transmitted to the brain in the form of material neural signals. If the mind were immaterial, then the information carried by these neural signals would have to be transformed into the immaterial form of the mind. The immaterial mind would also have to be able to cause material consequences via muscles. It is known that the brain controls muscles via material neural signals; thus the command information created by the immaterial mind would have to be converted into these material signals. However, a converter that transforms immaterial information into material processes is impossible as energy and/or matter would have to be created from nothing.

Cartesian dualism has also another difficult problem that actually makes the whole idea unscientific. The assumed immaterial mind would power cognition and consciousness and would be the agent that does the thinking and wanting. But, how would it do it? Is there inside the brain some immaterial entity, a homunculus, that does the thinking and perceives what the senses provide to it? How would this entity do it? Immaterial entities are, by definition, beyond the inspection powers of material means, that is, the material instruments we have or might have in the future. Therefore we cannot get any objective information about the assumed immaterial workings of the mind; any information that we might believe to have via introspection would be as good as any imagination. (Psychology does not tell anything about the actual enabling mechanisms of the mind. In the same way, you will not learn anything about the innards of a computer by inspecting the outward appearance and behavior of the executed programs.) Thus, the nature of the immaterial mind would remain in the twilight zones of imagination and make-believe, outside the realm of sound empirically testable science. By resorting to the immaterial mind, Cartesian dualism has explained what is to be explained by unexplainable; this is not science, this is not sound reasoning, this is not reasoning at all.

The conclusion: Cartesian dualism is wrong and immaterial minds do not exist. The mind appears as immaterial only because we do not perceive the underlying material machinery, the neurons and their firings. The brain does not observe the neural firing patterns and determine that this and that firing pattern would represent this and that entity. The brain does not inspect its neurons and their firings as such and does not naturally create mental images of these. (Now that we know that neurons exist, we can imagine them, but this is beside the point.)

2.4. Property Dualism

No matter how wrong Cartesian dualism would seem to be, we may still feel that there must be a fundamental difference between the body and the mind, the neural processes and the subjective experience. Could we somehow save the mind-body division while rejecting the immaterial substance? *Property dualism* tries to dot that.

Property dualism is a materialist theory that rejects the existence of an immaterial soul. There are no separate substances for the mind and body. The only substance is the physical material one with its material events. However, according to property dualism, certain material assemblies and events have two different kinds of properties, namely physical ones and mental ones. The mental properties emerge from the events of the physical matter, but *are not* reducible to those. According to property dualism, mental events supervene on material events, they "appear on top of these". By definition, a *supervenient* phenomenon S emerges from the material event M in such a way that whenever S is present, M must be present, too. Furthermore, S can change only if M changes. However, S does not have to be present every time M is present. In this respect, a supervenient phenomenon S can be compared to a shadow. S is like the shadow of M, present only when there is light and M is present, but absent, when M is present in darkness, Fig. 2.2.

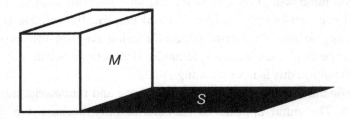

Fig. 2.2. A supervenient phenomenon S is like the shadow of M, present only when there is light and M is present, but absent, when M is present in darkness. When M changes, S changes too, but S cannot change on its own (illumination is constant).

What would be the mechanism of emergence here? Property dualism does not have a good answer, because it is maintained that emergent mental properties are not reducible to physical properties. Therefore, from the engineering point of view, emergence is not an explanation at all. At best it is a quasi-explanation, where an unexplainable mechanism, the emergence, is used to explain the original problem. Therefore it is difficult to see how property dualism would really explain anything.

2.5. The Identity Theory

The identity theory solves the problem of emergence by denying supervenient phenomena. According to the identity theory, mind is the action of nerve cells and their connections. Francis Crick summarises this in his book *Astonishing Hypothesis*: "Your sense of personal identity and free will are in fact no more than the behavior of a vast assembly of nerve cells" [Crick 1994].

There are some varieties of the identity theory. The *type-type identity* model proposes that a certain type of neural activity always corresponds to a certain type of mental activity and vice versa. The *token-token identity* model proposes that a certain type of mental activity may be generated by different types of neural activities. This would be just like in the computer, where, for instance, a certain kind of graphical interface may be generated by different algorithms. The token-token model could appeal to engineers as it would seem to allow the realization of the neural mechanisms by various ways, also by non-biological electronic ways.

The identity theory is a simple one, apparently working without any hypotheses about emergence or other quasi-explanations. As such, it would seem to be the one to be preferred. However, difficult problems remain. If the mind is the same as the neural processes of the brain, *then why are some neural processes perceived as mental content, while others are not?* Why is no mental content perceived during dreamless sleep, while many neural processes still continue in the brain? The identity theory does recognize this problem and proposes an answer: Those neural states that are identical with mental states, have an additional layer of mental properties. Somehow this proposition seems to weaken the whole point. It seems that instead of explaining the mind in terms of neural states, the identity theory is introducing another layer to be explained and in this way the searched explanation slips further away.

This problem has been observed by some philosophers. For instance, Maslin laments that the identity theory may escape the dualism of substances (with a very narrow margin perhaps), but not the dualism of properties [Maslin 2001]. This takes us back to the property dualism and, by a nasty surprise, the concept of emergence re-emerges.

It is an empirical fact that neural processes exist and are related to perceived mental content and therefore there may be some truth in the identity theory. It is also obvious that neural processes may appear as mental content only under certain conditions, but what would these conditions be? Property dualism and identity theories have not provided working answers.

2.6. The Real Problem of Consciousness

It could be easy to think that in this day and age the old dusty philosophical mind-body theories are no longer relevant in modern research, now that we have all the technology and instruments and real understanding of biology. Today, it is known that in the brain there are material neurons, synapses and glia, where physical processes take place and without these no consciousness, no subjective experience can occur. Nowadays these material neural processes can be externally detected by a number of methods and instruments and by these means some correlations between the neural processes and reported subjective experiences can be found. Therefore, should we just look for the neural areas in the brain that are correlated with conscious experience and when these areas, the neural correlates of consciousness (NCC), are found, declare the phenomenon of consciousness solved for good in the spirit of a modernized identity theory?

Some eminent researchers like Crick and Koch [1992] have investigated along this avenue. Surely, this work is important, because the workings of the neural machinery must be studied and understood before we can explain the workings of the brain. However, would this research explain also the appearance of the conscious mind? Does the observation that the activation of a certain group of neurons at the visual cortex is associated with the subjective experience of red, really explain, how this experience would arise? Not really. It is known that certain opioid peptides like endorphin affect synaptic transmission and cause euphoric effects. Does this explain the experience of feeling good; how, exactly, the releasing of opioids in the brain generates the *subjective feeling* of pleasure? No, not really. The crucial step in the explanation remains missing. There is nothing in these proposed explanations that would explain, why and how any subjective experience would arise from these processes. NCC theories may look shiny and modern, but their explanatory power is not any better than that of those old philosophical theories.

We can explain, how certain neuron groups work and we can explain, how certain opioids and other chemicals affect the neural synaptic transmission and we may speculate about some exotic (and very unlikely) quantum processes, but the real mind-body problem persists. We have not been able to explain, how these real or speculated physical processes could give rise to the inner subjective experience; a gap between the explanations of the physical processes and the explanation of the subjective experience remains. The existence of this *explanatory gap* has been recognized, for instance, by Levine [1983] and Chalmers [1995a, 1995b]. The old dusty philosophical mind-body theories are just about this explanatory gap in philosophical terms. We can explain the physical processes of the brain, but we cannot readily explain how they could give rise to the subjective experience.

Chalmers calls the explanatory gap the *hard problem of consciousness*. So hard this problem is that many contemporary texts (not this one) about machine consciousness cunningly by-pass or try to dissolve this issue altogether.

However, the explanatory gap offers another angle of approach to the problem of consciousness; this approach begins with the observation that our mind is not able to perceive the actual neural activity in the brain. We do not observe any neural firings, let alone the neurons themselves. Instead, we observe the vivid world around us and the feelings that our body provides for us; this is our subjective experience. Yet, it is for sure that all this information in the brain is in the form of neural activity. Therefore, *there must be a phenomenon, process or a system property that causes some neural activity to appear internally as the subjective experience.* The real mind-body problem relates to the founding out of this mechanism. But there is more.

The complete problem of consciousness includes the following questions:

1. How does *some* neural activity appear internally as the subjective experience?
2. How does the subject become aware of its mental content?
3. How does the impression of I, the perceiving, reasoning and wanting self, arise?

The solving of these problems constitutes the explanation of consciousness. The first question is the crucial one; any proper theory of consciousness must address it, otherwise the theory does not explain consciousness. The desired theory that solves these problems should be a practical one that would allow, at least in principle, the creation of conscious robots and artifacts; if we cannot reproduce it, we have not understood it. Thus, the explanation of consciousness should not be a dualist theory and it should not rely on the concepts of emergence and supervenience. The founding issues and proposed solutions of this kind of an approach are presented in the following chapters.

Summary

- The mind appears as immaterial.
- Cartesian dualism proposes that mind and body are separate immaterial and material substances.
- Cartesian dualism leads to the mind-body interaction problem.
- Property dualism denies the dualism of substances and proposes the dualism of properties instead.
- The identity theory proposes that the neural processes constitute the mind.
- The explanatory gap remains; how does the material brain produce the inner appearances of the external world and the material body and one's own thoughts? This is the hard problem of consciousness.

Chapter 3

Consciousness and Subjective Experience

3.1. Theories of Consciousness

We all are conscious beings, therefore we all should know what consciousness is. When we are conscious, we know that we are, but when we are not, we do not know that we are not conscious. Nevertheless, the exact definition of consciousness has been difficult and many attempted definitions turn out to be circular ones, explaining the entity to be explained by itself. For example, some dictionaries define consciousness as the state of being conscious.

According to folk psychology, we are conscious, when we feel pain and pleasure. We do things consciously, when the acts are a result of our intention and planning and we are able to report that we are doing and have done these acts. We are conscious, when we are able to observe the external world and the states of our body. We are conscious, when we have our silent inner speech and are aware of it. *Popular hallmarks of consciousness* like these can be easily listed, but these do not necessarily reveal much about the actual phenomenon of consciousness. Obviously, a good theory of consciousness should explain all everyday observations of consciousness, in one way or another. This theory should also explain all clinical findings about consciousness. But, most importantly, the theory should address the explanatory gap and explain, how the internal subjective experience could arise from the neural activity. This explanation should be based on physical and information theoretical facts and it should allow, at least in principle, the creation of conscious artifacts.

Recently many philosophical theories about consciousness have been proposed. None of these can be said to be complete, because they have not been able solve the explanatory gap. However, some theories have tried to solve the explanatory gap by *explaining it away*, without success. The explanatory gap may have been recognized, but nevertheless, in many cases philosophers and engineers have been erecting high-rising skyscrapers of attempted explanation on the both banks of the explanatory gap instead of building a bridge over the gorge.

Philosophical theories have not really offered a definite workable blueprint for conscious artifacts. This is typical to philosophical theories; a philosopher is not (usually) an engineer and therefore is not accustomed to the strict requirements of engineering practice.

A quick research into the consciousness literature reveals another conceptual problem; "consciousness" is used to refer to different aspects of mind and cognition. Sometimes "consciousness" is used to refer to the phenomenal aspect and sometimes it is used to refer to the *contents of consciousness* and *situation awareness*.

The contents of consciousness include sensory percepts, thoughts and emotional feelings. Situation awareness is related to the "who, where, when, what, why" information at each moment. A person with diminished situation awareness may not know who he is, where he is and what has happened; he has lost the access to this information in his mind and is not able to bring it into the contents of consciousness.

The behavior of the contents of consciousness and situation awareness can be explained by various associative and information integration processes and at first sight it might appear that the whole problem of consciousness were solved. However, the explanation of the processes of the contents of consciousness and situation awareness does not automatically provide a solution to the explanatory gap; the true problem, the arising of the phenomenal appearance of the contents of consciousness in the first place has not been explained.

Block [1995] has tried to clarify this situation by dividing the concept of consciousness into two components, which he calls the phenomenal (P) consciousness and the access (A) consciousness. P-consciousness relates to the raw perception of external and internal sensory information and feelings, while A-consciousness relates to the ability to reason, remember, report and control behavior. It is obvious that A-consciousness relates to the situation awareness and cognitive abilities of the conscious subject and consequently may be explained by the operation of information processing mechanisms. On the other hand, P-consciousness relates to the phenomenal aspect of consciousness, "how everything feels". Explaining P-consciousness would involve the explanation of feeling; what is feeling, why it feels like something and how a cognitive system can feel anything. Cognitive abilities may be externally assessed, but the subjective experience of feelings remain personal and can be observed only indirectly.

A-consciousness has function, as it is related to the cognitive abilities of the subject. P-consciousness may or may not have a function depending on its definition by different researchers. In order to clarify this situation, Boltuc [2009] has proposed an extended division of the concept of consciousness. According to this, P-consciousness is divided into H-consciousness or Hard consciousness and a new P-consciousness. H-consciousness is pure subjective phenomenal consciousness without any function and therefore it relates directly to the explanatory gap and the hard problem of Chalmers. The new P-consciousness involves the functional aspects of the original P-consciousness; being functional, these should be easy to explain. Thus, in effect, there would be two different concepts of consciousness; the hard H-consciousness and the easy functional F-consciousness. This functional consciousness would include the A-consciousness and the functional components of the original P-consciousness.

What is functional consciousness? When you put your finger in the candle flame, you will experience pain and will withdraw your hand quickly. The experience of pain is the phenomenal aspect and the withdrawal is the functional reaction. Pain will also teach us not to do again things that lead to pain, not to put our finger in the candle flame. Similar learning can be realized in a machine without any real *feel* of pain. *Artificial functional pain* (like in a trivial machine or a computer simulation) can be defined as the process that realizes all functional aspects of pain. In a similar way, *artificial functional consciousness* could be defined as a process that realizes the functions of consciousness.

But, are there any functions that consciousness would execute? By definition, Boltuc's functionless H-consciousness is the real subjective experience, without which there is no consciousness. The functions of consciousness would rather fall under the label of cognition and according to this division, no functionality remains for consciousness. However, *the way of information processing that produces the inner appearance of consciousness, most probably has functional benefits*, while consciousness as the phenomenal inner appearance does not have any function. Thus, there are no functions that could be attributed to consciousness and therefore functional consciousness does not exist. Consequently, artifacts that are claimed to be conscious, because they implement "functional consciousness" are not conscious at all.

According to Boltuc, the phenomenal H-consciousness is the real consciousness. Also Harnad and Scherzer [2007] propose that consciousness is phenomenal. According to them, to be conscious is to have phenomenal feel; consciousness is feeling, nothing more, nothing less. This feel has no causal powers. The difference between conscious and non-conscious acts is feeling; conscious acts are felt, while non-conscious acts are not. According to Harnad, every conscious percept involves feeling; a conscious subject feels pain, fear, hunger, etc. Also visual perception would involve feeling, it would feel like something to see blue.

This view can be challenged. For instance, the author finds visual perception totally neutral and without a slightest feel comparable to the actual feel of something, like a touch, temperature, pain or other bodily feel, yet the author finds himself to be visually conscious. It may be argued that the phenomenal qualities of visual perception (colors etc.) would constitute the feel, but this would stretch, twist and redefine the meaning of "feel" towards the point, where "feeling" and "consciousness" were synonyms; "consciousness is feeling". At that point the explanatory power of the concept "feeling" would be lost, because synonyms do not explain anything. For instance, if you do not know, what "a car" is and you are told that it is "an automobile", you will not be any wiser. Therefore "feeling" in its conventional meaning is not a sufficient constituent of consciousness. The spectrum of the phenomenal experience of consciousness involves more than feeling.

Scientific explanations are causal explanations; they propose mechanisms that cause the phenomenon that is to be explained. Boltuc, Harnad, the author and others have proposed that the real phenomenal consciousness does not have any function and has therefore no causal powers; consciousness is not an executive agent or substance. All the functions that consciousness might seem to have, are effected by the neural processes behind the conscious experience, not by the conscious experience itself. Does this lead to the conclusion that consciousness cannot be causally explained at all? At least Harnad [Harnad & Scherzer 2007] would seem to think so. According to Harnad, natural forces are causal, but without any feel. Therefore the feel of consciousness cannot be explained with a combination of the natural forces. Causal force explanation would call for a natural mental force that has feel, such as a telekinetic force. However, no telekinetic forces exist and therefore no causal explanation for the feel of consciousness can be provided in this way.

Many things may follow causally from others, yet these things do not have any causal powers by themselves, they are not able lead to any further consequences. For example, your next door neighbour may have a noisy party in the middle of the night. You may be utterly irritated and may go banging your wall for all of your worth, but alas, with no effect. Your behavior can be causally explained even though it itself had no causal powers. The situation with the phenomenal consciousness may be similar.

It is true that in physics, there is no equation that would derive the feeling of feel from the application of natural forces. And how could there be; the equations of physics describe numerically quantified entities and yield numerical values, not the quantified phenomena. No matter how you manipulate the equations, no feel will turn out. This is a wrong way to look for phenomenal feelings. For the explanation of any feel you must study self-observing systems that have subjective experience.

A conscious subject must have various cognitive skills, but in order to be truly conscious, it has to have some kind of a phenomenal subjective experience. The ultimate explanation of consciousness is not an explanation of cognitive skills, instead it would be the explanation of the phenomena related to the subjective experience.

3.2. The Subjective Experience

Look around. What do you see? What do you hear? You see your environment, which seems to appear to you directly as it is and you hear the sounds of the environment also apparently directly as they sound. All this facilitates your *situation awareness*, your understanding of what is around you and how all this relates to your immediate past and future. Usually all this happens without any obvious effort at all. You are immersed in the world and your perception of this is a part of your *subjective experience*. If you were in a dark, silent, secluded place and you close your eyes, you would still see something and hear something; you would see the visual noise patterns that your visual sensory system produces and you would hear the sonic hiss that your auditory sensory system produces. You might also perceive your breathing, heart beat and the warmth of your body. You would also feel how your body touches the environment. These would be a part of the residual *background feel* of the subjective experience that normally goes unnoticed, yet this background feel is with you wherever you go. Human subjective experience has also other components; emotional feelings and our thoughts in the form of imaginations and the silent inner speech that is usually present whenever we are awake.

We are aware of our subjective experience, it is *the contents of our consciousness*. We experience awareness whenever we are awake. During sleep, when we are dreaming we are aware of the dream, but not of our environment and the external world related situation awareness is missing.

The subjective experience depicts and is about something. This phenomenon is known in the philosophy of mind as *intentionality* and *aboutness*; the contents of the mind, the consciously perceived internal appearances, are about real or imaginary objects and conditions, *not* about the neural processes that carry them. According to this philosophy, consciousness is always about something, the contents of consciousness cannot be empty. Therefore, to be conscious is to be conscious of something. (Please note, this philosophical concept of intentionality should not be confused with the everyday meaning of intention; as for example, "I have the firm intention to read this book".)

3.3. The Internal Appearance of Neural Activity

The neural processes of the brain are material processes that can be detected and inspected by various instruments. The findings provided by these instruments show that the brain is active in certain ways, but they do not show directly, what the subject is experiencing. The mind, on the other hand, is not able to observe directly the neural processes of the brain. Instead, some neural activities appear internally as the *subjective experience* in the form of mental content; the seen world, heard sounds, smell, taste, temperature, feel, pain, pleasure, as well as internal imaginations, inner speech etc., see Fig. 3.1.

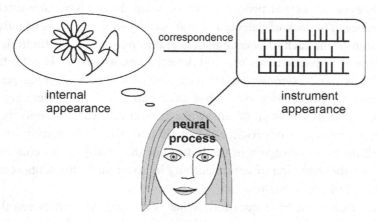

Fig. 3.1. Neural processes may be inspected by various instruments that produce instrument-specific data, "instrument appearances". Some, but not all, neural processes in the brain have an internal appearance; they depict something and the subject experiences these appearances consciously.

It is not proposed here that the subjective experience were something separate in addition to the related neural process; no, *the subjective experience is the internal appearance of certain neural processes*. It is not proposed either that the subjective experience of the external world were a representation or an image that would have to be inspected by the mind; no, the subjective experience is the result of the inspection of the world by the senses. *The mind does not internalize the perceived world, instead it externalizes the visual and auditory sensory percepts of the world, so that they appear as the external world.* The externalization of the information content of the internal neural firings and patterns is practically perfect. You do not inspect any internal projections of the external world, instead you inspect the external world directly; you look and the world is out there for you to see. Moreover, while doing this externalization, the mind associates various meanings with the objects of the external world. The internal appearance is internal, because it is available only to the subject itself and only available internally, even though this internal appearance may be that of an external outside world. The external world is shared by individuals, but the experience of the perceived appearances and evoked meanings is not.

However, all neural processes in the brain do not have any internal appearance. The lack of detected neural activity in the brain is usually an indication of the lack of consciousness and in extreme cases brain death. However, the reverse is not true; all detected neural activity is not related to and perceived as conscious subjective experience. A sleeping person does have neural activity even when the sleeper is not experiencing any dreams. Also in the brain of an awake person all neural activity is not related to conscious experience. Most (if not all) of the activity of the cerebellum (smaller region in the bottom of the brain) is sub-conscious. Therefore the detection of neural activity by external instruments is not a solid proof of consciousness.

This leads to another question: What is the difference between those neural activities that appear internally as subjective conscious experience and those neural activities that do not have that appearance? Why does the activity of certain neurons appear internally as subjective conscious experience while the activity of other, exactly similar neurons does not? Why does the activity of certain neurons appear at certain times as subjective conscious experience while at other times it does not? Indeed, what would be the difference between conscious and non-conscious neural operation as no special neurons for consciousness seem to exist? Solutions to these problems are discussed in the chapter 4, "Perception and Qualia".

Presently this problem cannot be solved directly with our brain imaging instruments. These instruments can create various "instrument appearances", symbolic representations such as graphics or data that depict some aspect of the brain activity, but the presence of any actual inner subjective experience remains beyond the resolving power of these instruments. An instrument may tell if there is brain activity, but it does not tell if this activity is perceived by the subject as an internal appearance. Some advances have been made, but it remains to be seen if we will have instruments that would reliably produce "instrument appearances" that would equal the internal appearances of the subject. An instrument like that might, for instance, display the dreams of a subject on a computer screen.

This problem has its counterpart in robotics. The internal activity of a robot "brain" can be inspected by using oscilloscopes, logic analyzers or other similar instruments. It is also possible to view the imagery and hear the sounds that the electronic processes of the robot brain are depicting. In spite of this, it cannot be seen directly that the robot is actually experiencing its electronic processes in the form of inner subjective experience. An oscilloscope alone cannot show that the robot is conscious, if we do not know a priori that the detected signals are definite indications of consciousness. Yet, it is the presence of the "internal appearance", the inner subjective experience that would make the very difference between a conscious and a non-conscious robot.

Consciousness is not directly related to intelligence; for instance, no higher intelligence is needed to feel pain consciously. Thus, the difference between a conscious and a non-conscious machine would not be the same as the difference between a dumb and a clever machine. However, it may be possible that true human-like intelligence cannot be realized without processes that also create the internal appearance of subjective experience, the presence of consciousness.

It might be argued that a robot with "an electronic brain" must necessarily have "internal appearances" because the electronic signals depict and are about something, Fig. 3.2. However, this is not necessarily the case, because also in the human brain there are neural activities that do not manifest themselves as inner appearance. It is possible that signals that are about something do not still create an internal depiction in the forms of conscious internal appearance and subjective experience. "Internal appearances" will not emerge automatically. Instead, some additional conditions are necessary. This issue is crucial to the understanding of consciousness and the creation of conscious robots. The mechanisms that may allow neural and artificial signals appear internally as subjective experience are discussed later on.

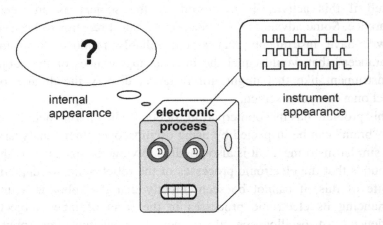

Fig. 3.2. The instrument appearance can prove that the robot has internal activity that is related, for instance, to sensory information. But the detected electronic activity alone is not a proof that this activity would appear internally as subjective experience. True conscious robots would have subjective experience.

Summary

- Certain physical processes of the brain appear *sometimes* internally as the conscious subjective experience.
- An explanation of cognition and the contents of consciousness does not constitute an explanation of the phenomenon of consciousness, why the neural activity has an internal appearance.
- The internal appearance does not execute any function, it is not an agent. Functional consciousness does not exist.
- The mind does not internalize the perceived world, instead it externalizes the visual and auditory sensory percepts of the world, so that they appear as the external world.

Chapter 4

Perception and Qualia

4.1. Perception and Recognition

4.1.1. *What is a Percept?*

Perception provides content to the conscious mind. The human mind acquires its experience about the environment and the body by the help of various sensory channels, such as the visual, auditory, touch, etc. Sensory systems use *receptors* or their technical counterparts, *sensors*, that are sensitive to the specific external stimuli of their kind. In the visual modality photons are sensed, in auditory modality vibrations of air pressure are sensed, in touch modality variations of contact pressure are sensed. Receptors and sensors are *transducers* that convert the external physical stimuli into the specific common form of physical representation that is used inside the system. In the brain this representation is in the form of neural signals, in the computer this representation is in the form of electric signals. Examples of artificial transducers are microphones, image sensors, pressure sensors, etc.

The mere act of *transduction* is not perception and a transduced signal is not a percept. The transduction of the sensory stimuli into a common form of representation is only the necessary first step towards perception. The information that the neural or electric signals contain must be differentiated and such forms of representation must be generated that allow the eventual *detection of content* and the *association of meaning*. This operation is called the *pre-processing* of the sensory information.

For example, a microphone does not understand what it hears, it does not perceive anything. A microphone only produces a varying electric voltage, which is an analog of the vibrations of the air pressure. As such it contains the sum effect of the individual sounds that are instantaneously present. For the eventual cognitive perception these individual sounds must be resolved in one way or another. In hearing the problem is the same; the eardrum vibrates to the sum of all sounds, which must be segregated. The inner ear provides the first step towards this by frequency analysis and in for that purpose the cochlea acts as the preprocessor for heard sounds. In computers the frequency analysis can be performed by filter banks or by Fourier transform. Image pre-processing involves the resolving of visual details and may include the detection of different colors and small-scale patterns.

An output signal from a pre-processor represents the presence of a certain resolved detail. A pattern of these signals may constitute the final perception of an object that is forwarded to the system. In many cases, however, the situation may not be so clear. A set of available details may be grouped in several ways that represent different objects and this leads to ambiguity. This ambiguity can be removed by the help of internal information from the system; the system may have learned that some patterns are more probable than others and represent frequently encountered sounds, objects and events. The utilization of internal information may be done by the modification of the pre-processed signal pattern by signals that come from system via reentry or feedback. These feedback signals may represent mental attention and context, experience and expectations. The process that combines and modulates sensory information with internal information is called *perception process*. The typical elements of perception are depicted in Fig. 4.1.

Here the *percept* is defined as follows: A single percept indicates the presence of a certain small-scale detail or feature. The ordered group of visual details make up a visual object. Likewise, temporally ordered tonal details make up a sound. The sensing of a complete object or entity leads to a large number of individual percepts that together constitute the full perception of the object, thus complete objects and entities are represented by patterns of percept signals. At the same time it is still possible to focus attention on any single detail.

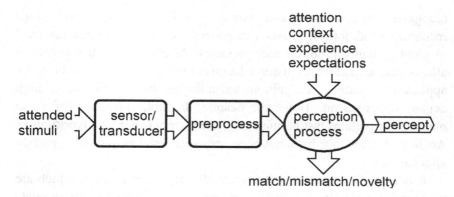

Fig. 4.1. Typical elements of perception. Perception is not a simple transduction or stimulus-response process, instead it is guided by attention and affected by feedback from the system in the form of context, experience and expectations. Match/mismatch/novelty conditions can be determined from the relationships between the sensed and expected signals.

The perception process is also a point, where the expected meet the actually sensed and the match and mismatch conditions between these can be detected. New percepts will not evoke any responses from the system and when no expectations are present then these new percepts should be deemed as novel.

4.1.2. *Is Perception the Same as Recognition?*

Is perception the same as recognition and is recognition the same as classification? If the answer were yes to both questions, then perception would be the same as classification and many problems of artificial cognition would be solved instantly. The perception process produces percepts of details and their combinations, which could be classified, named and thus recognized. Traditional artificial neural networks, such as the back-propagation networks, are classifiers and should therefore solve the problem of recognition. In practice this has not happened.

Artificial Intelligence has also pattern recognition methods that allow the recognition of visual, auditory and other signal patterns. These methods have been successfully used in vending machines that must recognize inserted bills and bottle recycling machines that must recognize valid bottles. In factories pattern recognition systems are used to identify faulty products and packages. Artificial speech recognition allows various automatic natural language telephone services. In many applications pattern recognition technologies have achieved a high degree of precision. However, despite the high precision of these methods, or perhaps just because of it, the pattern matching algorithms of Artificial Intelligence have not been very successful in general cognition applications.

It is useful to separate two aspects of recognition process, which are taken to be *continuity recognition* and *cognitive recognition*. These relate to all sensory modalities, but are especially prominent in visual perception.

In visual perception continuity recognition involves the ability to recognize a perceived object as the same, when the viewing distance, angle, motion or other aspect changes the appearance of the object. In this case the object does not have to be a familiar one. It is most useful to recognize a given object as the same even when it looks different at different positions and times.

Continuity recognition seems to be related to *change blindness*, the inability to notice certain changes in the perceived object during a changing situation.

Cognitive recognition involves the ability to evoke memorized experience that relates to the perceived object. We humans seem to recognize practically every object we see and most of the time we also recognize the words that we hear in our native language. At first sight we would seem to be master recognizes and consequently the process of perception would seem to be a process of recognition. However, the situation is not be that simple.

Figure 4.2 illustrates the function on human-style perception and recognition. A line drawing is recognized as a walking man. Exactly speaking, this figure is not taken as a real man, which it most certainly is not, instead it just reminds of a walking man. The simple lines of this drawing capture and represent some essential features that are present in the percept of a real walking man. This is typical to human perception. Instead of exact pattern classification, normalization and matching, the perception process only utilizes some essential features that are common to a number of similar cases. In the case of a man the clothing is not essential, as it changes. Likewise a man may be fat or thin, but it is not essential for the recognition, either. The essential features evoke associations that constitute the meaning for the perceived object. Thus, in cartoons the presentation of these bare bone essential features is sufficient; simple line drawings and caricatures are able to evoke ideas about the intended objects.

Fig. 4.2. What is here? A man walking with a document? Absolutely not. What you see are only some ink lines on the paper that may remind you of a walking man.

Thus, the incomplete percepts are able to evoke memories, which may include a name, a purpose, possibilities for action and emotional value. The evoked memories are context-related. For instance, we constantly see a large number of objects that do not evoke anything in our minds, because there is no need for that. Only those objects that are relevant to our current situation evoke memories that allow us to use those objects in the way that the situation calls for.

In this way, the associations evoked by the percepts allow us to perceive the possibilities for action that these percepts suggest to us instead of perceiving a collection of discrete entities without meaning. The world seems to offer us *"affordances"*, as proposed by Gibson [1966]. The perceived affordances are subjective products of the perceiving individual; different persons may see different possibilities.

Perception and imagination are connected. In many cases imagination must augment the limited amount of information that our senses are able to produce. We cannot see behind our backs and we cannot see behind objects, but we can imagine what is there and act accordingly. When we hear sounds, we can imagine the source of these. We can see what a person does at the moment and we can imagine what the person may do next.

Thus, perception is not the same as recognition and perception in cognitive systems should not be based on pre-programmed arrays of pattern matching modules. One might even say that objects are not really recognized. The percepts of objects only remind us of something; something that the objects might be, something that they might allow us to do. We "recognize" the perceived objects, because due to the general context we already "know" what they are. This kind of recognition is an associative process, where already minimal cues in the form of sensory features are usually sufficient.

4.2. Qualia

4.2.1. *What Are Qualia?*

Humans (and obviously animals) do not perceive neural percept signals as what they actually are; the firings of neurons. Instead, these neural firings appear as the qualities of the environment and its objects directly. The perceived qualities of the world and body sensations are called *qualia* and they constitute the appearance of the inner subjective experience.

The existence of qualia is directly related to the real mind-body problem: How can neural signal activity patterns in the brain appear internally as qualia instead of appearing as what they are; patterns of neural signals, or not appearing at all? This relates qualia to consciousness and the solving of this problem would constitute the solving of the first problem of consciousness. Therefore it is necessary to investigate the phenomenon of qualia more thoroughly.

The word "qualia" (singular quale) refers to the *phenomenal feel and quality* of sensory percepts. Typical qualia are, for example, the perceived colors, the timbre of sounds, the taste of edibles, the wetness of water, the softness of a tissue, the coldness of an ice cube, the feel of pain and pleasure. *Qualia are the qualities of percepts.*

Are qualia real world properties? Color television engineers know that when it comes to color qualia, this is not the case. Pure red and green colors seen together on the TV screen cause the perception of yellow, even though no yellow wavelength is present. This is a direct consequence of the color detection method of the eye; the utilization of three different receptors for three visual wavelength bands. Thus, the quale of yellow is generated by the visual sensory system. Also other evidence towards this conclusion exists. For instance, the visual after-image effect shows that qualia are indeed caused by the sensor itself.

It is also known that the artificial excitation of sensory nerves produces qualia. This fact is already utilized by the cochlear implant technology. A cochlear implant is an electronic device that is surgically implanted into the inner ear of a deaf person. This device receives sound via a microphone and stimulates the auditory nerves electrically according to the spectral content of the sound. This results in the perception of sound-like qualia and a patient with a cochlear implant will eventually understand speech. Here the auditory qualia are clearly generated by the artificial excitation of the auditory nerves.

Thus, *in general*, qualia are not some properties of the real world, instead they arise from the sensory organs' responses to stimuli. Nevertheless, qualia have a causal connection to the real world qualities, otherwise they would be useless. A certain real world property, for instance, a certain visible spectrum, will evoke a certain sensor output, which will appear with some consistency as a certain quale, a color.

The information conveyed by qualia can also be represented by indirect ways. This is demonstrated by Fig. 4.3, where the color information is presented by symbolic labels. More accurate color description could use numeric information that would indicate the value of hue and intensity of each pixel. This image shows an indirect representation of information by symbols that refer to some common knowledge. These symbols, labels or numbers, in itself do not contain any inherent information. (E.g. this could be demonstrated by using color names in some unknown language.) On the contrary, they must be interpreted by learned or otherwise available knowledge. This knowledge may be the meaning of color names or a common measure for color parameters etc.

Fig. 4.3. An image with indirect symbolic description of color information. In order to understand the color information one has to know what the labels mean.

The labels in the Fig. 4.3 tell what you would see if the picture were printed in full color. In fact, the labels describe the qualia that you would experience, and in this way, the labels are symbols for the corresponding qualia.

According to the previous example, the information that qualia seem to carry, can be conveyed also by other means. Jackson [1982] used this fact in his argument that qualia are not physical. All the information related to a percept can be described without qualia, therefore, according to Jackson, the inclusion of qualia does not add any physical information. Thus, qualia are not physical and therefore cannot be created by physical means. In his famous "Mary argument" Jackson presented a scientist, Mary, who knew everything about seeing colors; all the reactions in the eyes and in the brain, etc. However, Mary had been living in a black and white room for all her life and consequently had never actually seen any colors. However, she knew, what it would be like to see a color. Therefore the actual act of seeing of a color, the perceiving a color quale, would not increase her knowledge about seeing colors. Thus, qualia would not increase the amount of the physical information that Mary already had about colors, therefore qualia would not be physical.

Jackson's "Mary argument" is wrong. For the argument's sake, let us suppose that you are told, how it feels to put your finger in the candle flame. It will burn your finger badly, you will withdraw your hand immediately, your heartbeat will increase etc. In short, you would know exactly, what would happen. The actual execution of this act would be accompanied by qualia, but technically, you would not learn anything new that you did not already know. So, what would be the contribution of qualia if they did not increase your knowledge? Jackson argued that qualia are not physical, but actually the situation is the exact opposite. Information without qualia is a description; you have been told how it feels, but you have not actually felt it. *Qualia are physical experience*; you, your body and senses experience the situation and react to it, you actually feel and live through it.

If the picture of Fig. 4.3 were printed in color, color qualia could be directly and subjectively *experienced* requiring no further descriptions, symbols or commonly agreed comparison values. Red is red, sweet is sweet and pain is pain; we do not have to learn their names in order to experience them as they appear. We do not have to learn them via associations with some other entities. *Qualia are not symbols*, because they do not require any reference to some common knowledge in order to be understood. Thus, *in the brain, qualia are a direct way of experiencing information*. In other words, qualia are the subject's experience of the sensory stimuli; qualia are physical experience.

A further important conclusion can be drawn. Qualia are not perceived as the activity of neurons, because in the brain there are no receptors or sensors that could directly observe these. We do not perceive the spiking voltages of the neural signals that are carried by the neural network of the brain, but we perceive the effect that these signals have. There is no mystery here. *Qualia are the effects that are mediated by neural sensory signals*.

These are important issues that have profound implications for the research of machine consciousness.

4.2.2. *The Privacy of Qualia*

Qualia are subjective and available to others only via indirect description. We know the appearance and feel of our own qualia, but we do not know, if others have qualia and in general, what kind of qualities other people's qualia would have. We do not know, how other people perceive the world. However, the biological similarity of people leads us to assume that the qualia of other people are similar to ours, but this is not a proof.

The problem is illuminated by the condition of color blindness. Obviously colorblind people experience colors in a different way. For instance, they may perceive the colors red and green as the same color, but which color would that actually be? Would it be the red or the green of the people with the "normal" vision? Or would it be a some kind of brown? But, what would be the color of the "brown" of the colorblind people? And, would your "red" be similar to mine, for that matter? How can we know and does it really matter?

Previously it was stated that qualia arise from the sensory organs' responses to stimuli. These responses are in the form of neural signals and signal patterns. In principle these signals can be monitored by existing laboratory instruments, but in practice the skull of the subject would have to be opened for the access to individual neurons. There are other indirect non-invasive means of recording brain processes with various degrees of accuracy. These methods allow the detection of qualia-related electric and electromagnetic fields and potentials on the scalp and spatial metabolism changes inside the brain. The detected brain activity can be found to be associated with certain mental events by the subject's reports and some indirect ways. However, so far no qualia or any mental state has been detected directly. This is due to the way, in which existing brain imaging technologies and instruments detect physical processes in the brain. These instruments produce data and graphics that symbolize these processes. Also processes that involve qualia are represented in symbolic data.

Qualia are inside appearances in the brain and in order to see other people's qualia we should somehow connect their brains directly to ours, so that their qualia-generating brain activity would generate corresponding qualia in our brain. And yet, even in this case the generated qualia would be ours. There is no guarantee that these qualia would be even similar to the other person's qualia. The point is: *In general*, qualia are subjective and can only be experienced within the system itself and the brain seems to be the only instrument that can experience its own qualia.

4.2.3. *No Qualia, No Percepts*

The foregoing leads to the conclusion that in humans and obviously also in animals *qualia are, and constitute all direct, non-associated information about perceived objects and conditions.* There are no percepts without qualia and percepts are nothing more than qualia, when the associations that they may evoke are not included. This proposition can also be experimentally tested. For instance, image processing programs allow the step-by-step removal of image qualia, e.g. first the color qualia and then the grey-scale qualia. No direct perception of the image remains when all visual qualia are removed. A similar experiment can be executed with sound by using the tone and volume controls of an amplifier. First, the tonal qualities of the perceived sounds can be narrowed by the tone controls, then the intensity quale can be diminished until nothing is perceived. *Qualia are the way in which percepts appear to us.* Qualia are the qualities of percepts and they are also the way, in which different percepts differ from each other.

4.2.4. *Different Qualities of Qualia*

Seeing is different from hearing. Normally the percepts from the various sensory modalities do not get mixed; seeing something will not be perceived as a sound, touch or any other kind of percept. And, if it were, due to some physiological or medical reason (synesthesia, drugs), the percepts of sound and touch would still be different from any visual percepts. All sensory modalities produce percepts with subjective qualities, qualia, which are typical to the producing modality. Yet, in each case the experience is conveyed by neural signals that seem to be similar to each other. It is obvious that qualia must be different, how else could they depict the different qualities of the world. However, the question is: How can this differentiation be achieved by using similar neural signals?

In analog color television the colors are reconstructed by using three primary colors, red, green and blue (RGB). RGB circuits carry color information by three separate lines, one line for each color. The signals in these lines are superficially similar to each other and an external observer may not necessarily be able to identify each color by merely inspecting these signals. In this case the differentiation is realized by the hardware, the fixed wiring paths of these signals.

In the brain a similar scheme is used. The quality of the evoked qualia is tied to the identity of the nerves. Auditory qualia arise from the stimulation of the nerves that originate from the cochlea and terminate at the auditory cortex. Likewise, the stimulation of nerves that go to the visual cortex cause visual qualia, etc. This *specificity of nerves* can be demonstrated in various ways. For instance, a local push on the skin causes a tactile sensation while a similar push on the eye leads to the stimulation of visual nerves and causes visual sensations.

Thus, the evoked qualia seem to be determined by the fixed neural wiring. A certain sensory nerve would be associated with a certain quale. Obviously this is very much so, but this does not yet explain, what it is exactly that makes the quale seem different. A certain nerve carries by the variable pulse repetition rate the qualia of "reds", while the other one carries the qualia of "greens" for a given pixel in the retina. These qualia have a specific appearance, which is different from each other, but why? In technical terms fixed neural wiring would be sufficient for the discrimination of colors (and other sensory information), but the fixed wiring alone will not explain the appearance of the subjective experience.

This problem was recognized by Johannes Müller (1801–1858) and his theory known as *the Law of Qualities* tried to explain that [Gregory 1998]. According to that theory the quality of the evoked qualia is not generated at the origination points of the sensory nerves (the sensor), instead it is generated at the end points of these nerves, namely at the target area in the cortex. Each sensory modality has its target area in the cortex and this, according to Müller, is where the experience of qualia arises as a response to the stimuli carried by the incoming sensory nerves. Thus, the subjective experience of the quality of a quale would not be carried by the nerve signal, instead it would be created in the specific cortical areas for each sensory modality. Obviously these specific cortical areas would have to adapt to the type of information that they are receiving, like the serial auditory information and the spatially parallel visual information. The created qualia would have to reflect the format of the incoming information.

Balduzzi and Tononi [2009] have proposed that the quantity of consciousness is the amount of integrated information and qualia are specified by the set of informational relationships between the elements of the integrated information. Balduzzi and Tononi try to describe geometrically the entire set of informational relationships of the integrated information by considering a qualia space with an axis for each possible state of the complex. In this way each quale can, in principle, be mapped geometrically and a corresponding neural activity pattern may, perhaps, be found. However, this approach does not explain how the neural activity would give rise to any inner appearance of qualia in the first place; the actual problem of qualia is neither touched nor solved.

The Law of Qualities and information integration theories, if true, would invalidate *externalist theories of mind* [e.g. Manzotti & Tagliasco 2007] that state that the mind extends beyond the boundaries of the body. Thus, the external world with its properties could, in a way, be inside the mind while being outside the body and thus qualia would actually be external world properties. Manzotti and Tagliasco propose a kind of direct realism based on the identity between the physical process embedding the perceived object and the phenomenal experience itself; according to them, the act of observation, the observer and the observed entity cannot be split.

However, questions remain. Why would the cortex create consciously perceived qualia at all, because technically these would be redundant; all necessary information would already be present due to the hardwiring of the neural signal lines, just like in the color television. And secondly, what would be the actual mechanism that would generate the wide spectrum of qualia? The mere statement that says that it is here (at the target area in the cortex), where it happens, is not an explanation.

For a better understanding of the situation, the nature of the sensory information must be considered.

4.2.5. *Amodal Qualia*

It was stated before that, *in general*, qualia are not some properties of the real world. This statement needs to be clarified here. There are certain detectable features that are the same in several sensory modalities, such as the seen and heard direction of a sound source, felt and seen shapes and, especially, rhythm and interval duration. For example, feeling the teeth with the tip of the tongue results in shape percepts that are image-like. These properties are called *amodal features*. It is argued that these, when perceived, appear as *amodal qualia*. (This is not to be confused with the so-called amodal perception, e.g. the virtual perception of a partly hidden object as a whole.) In fact, amodal features are *invariants*; they are the same in the observed world and in the resulting neural activity. For instance, the rhythm of a piece of music is a feature of the actual sensed phenomenon and it should also be a feature of the resulting neural activity. Also changes in percepts, like spatial and temporal intensity changes etc. can be depicted in amodal ways. Please note that it is not claimed here that amodal *qualia* were properties of the external world. No, the external world properties that lead to amodal qualia are certain amodal features of the external world. Qualia are the internal appearance of the evoked neural activity.

It was also stated before that, *in general*, we cannot know what kind of qualities other people's qualia would have. However, amodal qualia may be an exception to this rule. For instance, rhythm is the same for everybody and can be considered as a *shared quale*. (Dancing would be a method for sharing rhythmic motional qualia.) In the case of cognitive robots this leads to an interesting result: We could know how an amodal quale would appear in a robot, if the robot had internal appearances in the first place. Nagel [1974] argued that we cannot possibly know how it feels to be a bat. However, it would seem that the consideration of amodal qualia could allow us to know to a certain extent how it feels to be a bat. Nevertheless, with this information only it is not possible to know that the bat would actually feel anything.

All features are not fundamentally amodal. Examples of features that are seemingly not amodal include colors, taste, smell, etc. However, these features may appear in the context of amodal features.

Amodal features do not depend on the material means that carry them. Thus a mechanism that carries amodal features may be compared to that of radio transmission, where the information is modulated on a carrier wave. The changes in the carrier wave reflect the changes in the modulating audio signal. The listener will become aware of the audio signal only, while the carrier wave remains hidden.

Figure 4.4 depicts the sensing of amodal features. In a system that conserves amodal features, a sensed amodal world property leads to the corresponding amodal sensory response. This response leads then to amodal sensory signals, which may lead to an amodal system reaction. Rhythm music can be considered as an example: Drum beat evokes beating neural signals and these in turn may lead to feet tapping to the beat. The conserved amodal quale is in this case the rhythm.

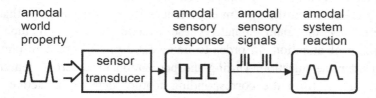

Fig. 4.4. In amodal feature conserving systems an amodal world property leads to the corresponding amodal sensory response, this leads to amodal sensory signals, which may lead to an amodal system reaction. An example: Drum beat – beating neural signals – feet tapping to the beat.

Amodal features offer also a natural way of sensorimotor integration. An example of sensorimotor integration is dance, where the rhythm of the music is also the rhythm of the motion. It is obvious that the perceived amodal feature, the rhythm, can be directly used as motor control commands. Also the production of speech would seem to utilize amodal features.

Obviously amodal qualia originate from an external source and are not a product of a cortical area. Does this then prove that in the case of amodal qualia the Law of Qualities is wrong and externalist theories are right and in other cases the Law of Qualities would be right and the externalist theories would be wrong? A theory that could explain all cases of a phenomenon would always be the desired one; do we have here a such theory?

The *system reactions approach to qualia* combines the external and internal aspects of qualia creation. According to this approach, in each case the special qualities of qualia arise from the system reactions that are triggered by the sensory signals. Amodal qualia would arise as a result of amodal sensory responses and other qualia would arise from non-amodal system reactions. Extreme examples of these kinds of qualia would be the feel of pain and pleasure.

Human body is equipped with a large number of pain receptors, which are actually nerve ends that sense cell damage. During a cell damage neural signals are transmitted to the cortex and the feeling of pain is created. The neural signal itself is similar to the other neural signals, which do not appear painful. Where does the hurting feeling come from? The author has proposed [Haikonen 2003] that the hurting feeling arises when the system experiences its system reactions that are triggered by the pain signal. These system reactions include the disruption of attention. In order to understand this mechanism one has to consider the question: During the perception of a certain feature, a feature signal from the corresponding feature detector is active. The presence of this signal indicates the presence of the corresponding feature, but what else might be perceived about this signal? A simple example illuminates this case.

Imagine listening to your favorite music. Imagine then, that due to some technical problem the music is intermittently switched on and off in a staccato style; after each fraction of second of music there would be an equal period of silence. How would that feel? You would be very irritated, at least. Here you would not only perceive the music, but also the on/off switching of the same; the perception of the music facilitates the perception of the on/off switching, no specific sensors are required. Consider now that the periodic on/off switching of the music would not take place outside your head, instead it would happen in your brain by the switching on and off the corresponding neural auditory signals. It is proposed that you would perceive the situation in a similar irritating way. In addition to the perceived music, you would also perceive the neural signal on/off modulation, which would have an irritating effect. It is proposed that the feeling of pain is related to a similar effect caused by another fast intermittent neural signal that would disrupt the attention on the ongoing activities.

Pain is a feeling that demands attention; something must be done to make the pain to stop. In order to get attention, pain would have to suppress other active signals, but not completely, because it would not be safe to make the subject unable to think and do anything at all. Therefore a periodic on/off switching or attenuation of other neural signals would be more useful. The author has proposed [Haikonen 2003] that this is exactly how pain works; it disrupts neural signals in various periodical and chaotic ways and these disruptions appear as the various feelings of pain. The on/off switching is a system reaction, which is observable within the system without any specific sensors as explained above and this appears as the quale of pain.

4.2.6. *Externalization, the Apparent Location of Percepts*

Natural visual perception creates an interesting illusion. The lens of the eye projects an image of the scene in front of it on the retina. Each point of this image stimulates the corresponding light-sensitive receptors on the retina according to the image's local luminous intensity. A large number of fibers transmit the neural output signals from these receptors to the brain, where the information appears internally as visual qualia.

The illusion is: This information originates from the receptors at the retina, yet the internal appearance is that of the scene and its objects that are out there, not on the retina. We do not see the retinal image as a map or a depiction of the outside, instead we have the impression of seeing the outside world directly as it is.

Moreover, the direct visual impression of the outside world is seamlessly connected to our ability to move; we can reach out and touch the objects that we see effortlessly without any apparent calculations of the required motion trajectories. All this information is in the brain in the form of neural activity, yet this *internal depiction is externalized*, the objects of the world and their locations appear to be out there.

Similar illusion of externalization occurs in the auditory perception. Sound is received by the ears, where the cochlea transforms it into neural signal trains. This is the actual origination point of the neural sound signals, yet the perceived location of the sounds appears to be out there.

In the brain, there is nothing in the sensory signals that would tell directly the origination point of the information. The neural fibers originate from the sensors, but the neural firings that they carry do not convey any information about their origination point. The neural fibers are not labeled and the firings are not either; they have only causal connection to the external stimuli. The strength of a stimulus may be encoded into the neural signal, but it does not tell if it is caused by a weak stimulus nearby or a far away strong stimulus that appears as a weak signal at that distance. Therefore, without any additional information, the sensory signal patterns would stand for stimuli with undetermined location.

The additional information for the determination of the location of external stimuli may be gained via explorative acts. A practical experiment illustrates the situation.

A simple sound externalization experiment may be executed easily with the aid of good quality headphones, a stereo headphone amplifier and a pair of omnidirectional microphones. The headphones must be of the closed design so that no external sound can disturb the experiment. In this experiment the perceived location of heard sounds are inside the head, but after an explorative act the sounds become externalized. The experiment is executed in three steps, see Fig. 4.5.

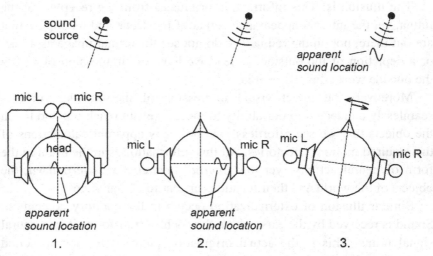

Fig. 4.5. The sound externalization experiment. 1. The left and right microphones are held together creating an apparent sound location inside the head. 2. Microphones near the ears create a transparent system, yet the apparent sound location remains inside the head. 3. The turning of the head resolves the situation and the sound is externalized.

The sound externalization experiment is executed in three steps:

1. The microphones are placed close to each other in front of the test subject's head. Obviously both microphones receive the same sound intensity and consequently the left and right speakers of the headphones generate a sound with equal intensity. The apparent sound location will be inside, in the middle of the head.

2. The head is kept still and the left and right microphones are brought next to the corresponding headphone speakers. (It is important that the headphones are of the closed design, otherwise acoustic feedback, "howling", can take place!) Now the situation corresponds to binaural listening. Good quality microphones, microphone amplifiers and headphones should become transparent and the test subject should hear as well as when no headphones were present. Thus, it could be expected that the sounds would appear to come from locations outside the head. However, this does not happen; the apparent sound locations will still remain inside the head.

3. The test subject turns his head. Suddenly, the sounds are externalized; they appear to come from outside and the headphones seem to go silent. Thereafter the perceived origination points of the sound stay outside even when the head is kept still and it will be possible to turn the head towards a sound source at will. (The vestibular apparatus of the inner ear and the neck muscle tension receptors detect the changes in the orientation of the head, therefore no visual cues are necessary and this experiment works also in total darkness.)

The lesson from this experiment is: Static perception does not convey enough information for the externalization of sounds; the sound sources do not appear to be out there. In this experiment the turning of the head and the associated change of the relative left and right speaker intensities and delays provide the additional information that gives unambiguous directions for the sounds. *A sound is externalized and seems to be out there, because a direction is seamlessly associated with it.*

In humans, sound direction detection is based on the intensity difference (at higher frequencies) and the time delay difference (at frequencies below 1500 Hz) between the ears and to a very small extent to the directional frequency filtering of the outer ear (pinna). The direction information provided by these means is, however, ambivalent, because a sound source in front of or behind the subject would generate a similar percept. The turning of the head resolves this ambivalence. Obviously in this experiment the pinnae have no influence as they are covered by the headphones.

This experiment highlights the importance of active exploration in perception. The turning of the head is an explorative act that provides the required additional information about the sound source direction in respect to the head.

Similar explorative acts are used in vision. The turning of the eyes provide direction information about the seen objects. However, when you keep your head still and turn only your eyes, you will notice that the world seems to be static, it is as if you were inspecting a photograph. The turning of the eyes does not provide information about the relative distances of the objects; for this purpose you must turn your head. This will change your vantage point slightly and you will notice that the objects seem to move a little bit in relation to each other, the visual scene is no longer static. You will see which objects are behind others. (These effects have nothing to do with neurology, they follow directly from the geometrical setup.) Additional explorative experiments are possible; the closing of eyes or covering them with hands will show that the visual world is out there, behind the eyelids or hands.

Thus, the perception of the location of visual and auditory objects is an active inspection and exploration process that involves physical adjustments of the sensors, like eye and head motions and hand motions (touch). These motions allow the acquisition of additional information that makes sense only if the perceived object is out there; other impressions would be conflicting each other. This impression of externalized objects and sound locations is most useful, but at the same time it also amplifies the illusion that qualia were real properties of the external world.

The externalization of sensory information applies also to body and contact sensors. When we touch something with our fingers, the sensation appears to take place at our finger tips, yet the neural activity that conveys this impression is located inside the brain. All the sensations that are generated by the various receptors in the skin, have an apparent location, which is the location of the corresponding receptor. Also pain may have an apparent location somewhere in the body, even though the actual feel of pain is created inside the brain. The receptor nerves do not have any intrinsic information about their points of origin and cannot therefore provide enough information for the externalization of their signals. The apparent locations of body sensations and pain outside the brain are related to the internal *body image*, which is created by the combination of information from various sensory modalities during experimentation in early childhood.

The created body image is not as robust as we might believe. The *rubber hand experiment* [Botvinick & Cohen 1998] and other similar experiments seem to verify that the locations of touch sensations are not intrinsically mediated by the touch nerves, because the perceived location can be quite easily changed. In a typical rubber hand experiment, a rubber hand replica is placed on the table in front of the test subject. The subject's own hand is kept below the table, so that the subject cannot see it. During the test the subject's own hand and the rubber hand are stroked synchronously with paint brushes. The subject sees that the rubber hand is being stroked and feels the coincident stroking in his own hand. After a short while the subject may begin to feel that the touch-sensation originates from the rubber hand; the subject's body image has been distorted. The rubber hand experiment shows that the body image is indeed created by explorative acts and is not an inherent fixed one.

In general, body image information can be achieved by the cooperation of various senses, especially by seeing and touching various parts of the body. A pain sensation without a definite position may be located by touching and feeling the affected body area and noting where the touch sensation and the original pain sensation coincide; "feeling where it hurts". Due to the acquired body image we readily know, for instance, which nerves report conditions in our fingers and when we cut a finger, the pain will not seem to be in the brain, where it is actually generated. Instead, the pain appears right there in the finger and will move around with the finger when we move our hand.

Yet, there may be cases where the body image is not perfect and we must touch and feel, in order to find out where it actually hurts. The externalization of a contact sensor sensation is based on exploration and the resulting association of body part positions with it.

Summary

- A percept signal originates from a sensor or from a sensory preprocessor.
- A percept signal depicts a sensed feature.
- Qualia are the qualities of percepts; the way in which consciously sensed sensory information appears in the mind.
- Qualia are direct and require no interpretation or associative learning.
- Qualia constitute all direct and primary, non-associated information in any percept.
- Amodal qualia depict invariant qualities.
- Amodal features, like rhythm, should appear internally as the perceived qualia.
- Qualia are subjective and available to others only via indirect description. Amodal qualia may be an exception.
- Sensory qualia are externalized; their apparent location is outside the brain.
- Body image is related to the externalization of body sensations.

Chapter 5

From Perception to Consciousness

5.1. No Percepts – No Consciousness

Introspection shows that in the mind every sensory percept seems to appear as some kind of a quale, at least a weak one. Practical experiments seem to show that the loss of qualia leads to the loss of percepts. This fits well with the idea that qualia constitute all direct and primary, non-associated information in any percept. Therefore, qualia seem to be the first and primary way in which sensory information appears in mind; percepts appear internally as qualia.

In addition to the sensory based content, the mind has also internally generated content such as emotional feelings, thoughts and imagination. Introspection shows that also these seem to appear as qualia. Verbal thoughts may be about abstract matters, but their carriers are words, which are nevertheless perceived as kinds of heard sound patterns with the corresponding auditory qualia. Imaginations are usually perceived in the form of vague images and therefore they have visual qualia.

Consciousness is based on perception. To be consciously aware of something is to perceive it, as a real world entity or as an imagined one. When perception processes cease, nothing will be perceived. All percepts, including the residual ones, are lost and the conscious experience vanishes, as there is nothing to be conscious of any more.

To perceive something is to experience the corresponding qualia, because qualia are the very way in which percepts manifest themselves. Therefore, when qualia are lost, percepts are lost. This connects qualia to consciousness. This connection is demonstrated by the application of anesthesia. Full clinical anesthesia is designed to remove the feel of pain and also the experience and memory of the otherwise traumatic operation. Successful anesthesia does this by removing the patient's consciousness. Pain is a quale, thus the application of anesthesia shows that the removal of consciousness removes also qualia.

This observation leads to the question: There are no qualia without consciousness, but can there be consciousness without qualia? If not, can consciousness and the experience of qualia be equated; does the presence and flow of qualia actually constitute the phenomenon that is called consciousness?

To be conscious is to have subjective inner experience. The inner experience consists of percepts and these in turn have the appearance of qualia. If qualia are removed, then percepts disappear and the inner experience becomes void. To be conscious is to be conscious of something; there is no consciousness without content. Thus, *consciousness is the presence and flow of qualia.*

There is no human-style consciousness without qualia, and machines with human-style consciousness would have to have some kind of qualia. However, machine qualia do not have to be exactly similar to human qualia. Nevertheless, a conscious machine would have to perceive the world just like we do in a direct way, as opposed to symbolic representations calling for interpretation. A conscious robot must be able to externalize non-contact sensory percepts and it must also be able to build a body image with the information from contact sensors.

It can be argued that machines might be conscious in some other way, but in that case we would not be talking about real human-like phenomenal consciousness and that would be another story.

5.2. Attention and Consciousness

Attention is the general name for mental selection processes. There are several varieties of attention. *Sensory attention* focuses the perception process on the selected objects. *Inner attention* focuses recall and thought processes on the selected topic.

Cognitive systems need various mechanisms of attention. Attention mechanisms are necessary in sensory systems, which produce constantly large quantities of information. Only a small part of the available information is usually relevant and calls for further processing. The rest of the sensory information must not be totally ignored, but its full scale processing would unnecessarily deplete the processing power of the system. Attention selects the items that are allowed into the working memory areas of the brain. The working memory areas have limited capacity and this limits the number of items that can be simultaneously in the focus of conscious attention.

Attention is also needed as a *pinpointing device*. Associative learning, such as the learning of the names of objects call for the pinpointing of the entities to be associated with each other. This association takes place via the temporal coincidence of the percepts of the entities to be associated. Usually there are several entities present at any given time, therefore pinpointing is necessary in order to avoid unnecessary or even harmful associations.

The attention process has three phases: 1. engage attention, 2. sustain attention, 3. disengage attention. *Attention span* is the period of time that attention on a given topic can be sustained. *Attention deficiency* is a disorder that affects the sustaining of attention so that the subject is easily distracted and has difficulties in maintaining focus on the on-going task. There are also certain mental disorders that make it difficult to disengage and refocus attention.

Sensory attention may be engaged by novelty, change, intensity, emotional significance, associated meaning or context (related to needs etc.) For instance, auditory attention may be captured by a loud or novel sound or a sounds with emotional significance, such as own name, pleasantness or threat. Visual attention may be captured by an appearance that differs strongly from the background, by moving or changing visual objects, or by objects that have strong emotional significance (e.g. threatening objects or objects of desire). Sensory attention process calls for the minimal processing of the unattended information, too, because otherwise sensory attention could not be refocused by the above phenomena. This is demonstrated, for instance, by the so-called *cocktail party effect*; it is possible to attend to only one talker under the noise of the party and yet it is possible to shift attention immediately to another stream of speech if some significant word, like one's own name, is uttered [Nairne 1997].

Inner attention is controlled by sensory factors and internal factors such as context, emotional states and motivation. The possible experience of pain and pleasure affect the sustaining of inner attention. Pain tries to disrupt inner attention, while pleasure tries to sustain it and in this way also the activity that is producing the feel of pleasure.

Attention and consciousness are related. Attention mechanisms bring objects into the perceptual short-term working memory areas, from where they are broadcast for further cognitive processing and may be memorized. On the other hand, it was argued afore that the contents of the perceptual short-term working memories, percepts, are also the contents of consciousness. Thus, attention mechanisms and short-term memories would seem to be technical prerequisites for consciousness. These short-term memories may not necessarily operate in a similar way to the addressable digital computer memories, but may rather be circuits that sustain their activity for a while.

5.3. The Difference Between Conscious and Non-Conscious Perception

In the foregoing it was argued that qualia are the way in which information manifests itself in the mind of a conscious subject and to be conscious is to have qualia-based inner subjective experience. Obviously, to have qualia is to be aware of them. For instance, you are not in pain if you are not aware of it. To have qualia is to experience the information and therefore qualia would seem to be inherently conscious, but there are still further questions to be answered.

Qualia are related to conscious percepts; they are the way in which conscious percepts appear to us. However, it is known that sub-conscious sensory information without the appearance of consciously perceived qualia, may nevertheless affect our mind and behavior. Therefore, the question remains: What makes the subject to become aware of its percepts; what would be the difference between conscious and non-conscious perception?

The simple answer to the above question would be reportability. If the subject is not able to report a percept, then obviously the percept has been sub-conscious (or non-existent). But there is more. Let's take an example. This happens when I hear a sound consciously:

- I can hear it and I can focus my attention on it (qualia and attention)
- I can report that I hear it (report)
- I can remember hearing it (memory making)
- I can turn my head towards its source (sensorimotor information integration)
- I can associate some meaning with it (information integration; learning and evocation of associated meaning; the sound of a bell, etc.)
- I can report the associated meaning

If a sound is only heard fleetingly and no report can be generated to oneself, then the perception of that sound has not taken place consciously. Fully conscious perception leads to a number of consequences that are facilitated by a large number of cross-connections; the activation of associative links and the evocation of responses. It is known that the brain requires a certain minimum duration for a sensory stimulus in order to perceive it consciously. A sensory stimulus that is too short, will not capture wide attention; it is not able to activate associative links before dying out and therefore only limited and short-lived memory traces will be generated, evoked responses will be minimal or non-existent and no report can be given. Sensorimotor integration will be minimal. There will be minimal or non-existent system reactions. Without system reactions there will be no such qualia as pain and pleasure. Without amodal system reactions there will be no amodal qualia.

The above observations apply to conscious perception in general. The difference between conscious and non-conscious perception is the difference between *information integration*; the number of activated connections that enable the *reporting, recalling, evoking of associated meanings and response generation.*

5.4. Information Integration and Consciousness

Some researchers have proposed that conscious states are related to the active cross-connections between the various parts of the brain and also the author has recognized this connection. Therefore, would information integration alone be able to generate consciousness?

There are some proposals towards that direction. The Information Integration Theory of Tononi proposes that consciousness corresponds to the brain's (or other cognitive system's) capacity to integrate information [Tononi 2004]. Von der Malsburg [1997] has also recognized the role of cross-connections and integrated activity in consciousness. Also experimental evidence seems to point towards this conclusion. For instance, the research of Massimi *et al.* [2005] seems to show that the active neural communication between the different parts of the cerebral cortex were related to consciousness; during dreamless sleep this communication breaks down and consciousness as well as the inner speech vanishes. Likewise, the function of information integration via the focusing of attention on the same topic by "broadcasts" is an essential feature in the cognitive models of Baars [1988] and Shanahan [2010]. In their models this activity is the one that is said to make the distinction between conscious and non-conscious processing.

As stated before, it is obvious that active cross-connections between the various parts of the brain (or an artificial cognitive system) are necessary for the realization of one feature of consciousness, namely that of reportability; conscious states are reportable states, internally and externally. Without the cross-connections and the exchanges of broadcasts between the various brain areas no reports can be transmitted across the brain and no associative memories of the event can be created. Without these cross-connections and internal reports we could not even reach out and touch objects that we see. An event that does not have any effects and cannot be remembered even for a little while is hardly recognized as a conscious one.

But, make no mistake here. The mere number of active connections or the quantity of "information integration" no matter how large, does not make the system conscious. The primary criterion for consciousness is still the requirement of the internal appearance in the form of qualia.

The requirement of information integration is secondary. Cross-connections and information integration are related to consciousness, because they facilitate reportability, memory making, situation awareness and many cognitive functions as well. However, they are not able to generate consciousness alone; the conditions that allow the internal appearances in the first place must be present. Information integration alone does not explain consciousness, because it does not address the explanatory gap of qualia; it is only related to some secondary aspects of consciousness.

5.5. What is Consciousness?

According to the previous chapters, a conscious agent differs from non-conscious agents in one important respect; it has subjective experience with qualia-based internal appearance. Thus, consciousness can be defined as qualia based perception, that is, the presence of the phenomenal subjective experience; the internal appearance of the perception-related neural activity. The contents of consciousness consists of the direct, externalized sensory percepts of the environment and also the perception of the body, mental perception of thoughts, imaginations, emotions, the feel of pain and pleasure etc. and nothing else. A similar idea has been proposed by David Hume (1711–1776): "I can observe nothing but perception" [Hume 2000].

To be conscious is to have the internal appearance. To explain consciousness is to explain the mechanism that facilitates this internal appearance. Neurons are quite similar in humans and animals. Therefore it may be assumed that also animals are conscious, because most probably they also have the mechanism for the internal appearance. However, their contents of consciousness is not necessarily the same.

Consciousness as the phenomenal internal appearance should not be confused with the contents of consciousness and cognition. The mechanism for consciousness may be the same for different agents, but the contents of consciousness and the cognitive abilities may not be. Many proposed explanations for consciousness are actually attempted explanations for the contents of consciousness and cognitive functions.

No matter how perfect these might be, they are not explanations of consciousness. They do not cross the explanatory gap.

Consciousness should not be confused with computation, either. Computation is the execution of algorithms and as such has nothing to do with phenomenal internal appearances and experience.

The view of consciousness as the presence of internal phenomenal subjective experience is shared, for instance, by Boltuc [2009] and Gamez [2008], who has stated: "Consciousness is the presence of a phenomenal world". This proposition is also seconded by Aleksander [2009]. Also Harnad has emphasized the phenomenal aspect: "Consciousness is feeling" [Harnad & Scherzer 2007].

Consciousness defined in this way is not an agent, an immaterial one or based on some exotic substance. The presence of a qualia-based internal appearance is only a special way of *internal manifestation of information* and as such does not have any executive or cognitive powers. The causal cognitive powers arise from the neural information processing ways that also give rise to the inner subjective experience.

The inner subjective experience constitutes the contents of consciousness and is always about something. To be conscious is to be conscious of something. The contents of consciousness cannot be empty, because to be conscious of nothing is not a conscious state at all.

It is a fact that that the brain operates with neural processes and it is a fact that we experience some of these as subjective appearances. The fitting of these two facts together constitutes the explanation of consciousness.

Summary

- Consciousness is the phenomenal internal appearance, "subjective experience" of perception-related neural activity.
- Qualia are the qualities of conscious percepts.
- A conscious episode involves perception, attention, short-term memory, information integration and report.
- Situation awareness, the contents of consciousness and cognition should not be confused with phenomenal consciousness.
- Consciousness is not computational.

Chapter 6

Emotions and Consciousness

6.1. Emotions and Feelings

Human conscious experience is colored with emotions. Strong emotional feelings make you feel alive and this is what consciousness is ultimately supposed to be about; to have the feeling of being alive. Therefore, can we feel to be alive without emotions and are emotions necessary for consciousness?

Should robots also feel to be alive and have emotions then? Or, should humans be better off without emotions; we do have plenteous examples of bad outcomes also in grand scales caused by emotional rage and attitudes. On the other hand, some emotions would seem to be good for us; therefore, should we try to implement useful emotions in a robot? Or, would it be possible that conscious robots would behave in emotional ways even if those were not explicitly implemented in their design? There are good reasons to consider emotions and emotional feelings also in the context of conscious robots.

No exact definition of emotional states exists, but a general idea may be given by a list of typical emotions:

Anger; Curiosity; Boredom; Desire; Despair; Disappointment; Disgust; Empathy; Envy; Fear; Gratitude; Grief; Guilt; Happiness; Hatred; Hope; Horror; Hostility; Jealousy; Love; Lust; Pity; Pleasure; Pride; Remorse; Sadness; Shame; Shyness; Sorrow; Surprise...

It can be seen that emotions are reactions to everyday situations in life and in many cases they contain the element of conflict. Indeed, emotions are triggered by certain situations and they involve consciously perceived *arousal, subjective feeling* and *physiological symptoms.* The reason for a specific emotion is, however, not always consciously perceived.

The physiological symptoms may include some of the following:

Facial expressions; crying; smiling; laughter; blushing; sweating; paleness; shivering; vocal modulation; changes in heart rate; changes in breathing rate; changes in blood pressure; stomach symptoms...

Many of the physiological symptoms of emotions would seem to be useless or even harmful. Why should a shy person have to sweat or feel sick when delivering a speech to a large audience? Would these kinds of emotional reactions be useful to humanoid robots? A shy clumsy robot fearing to talk might appear funny, but not very useful. Obviously there are some evolutionary reasons for the physiological symptoms of emotions, but these reasons might arise from the unplanned evolution of the central nervous system and its connections to the homeostatic control of the body and not so much from any survival aspect.

Facial expressions of emotions may be useful, because these can be used to communicate emotional states to others so that they can adjust their behavior accordingly.

6.2. The Qualia of Emotions

Emotions are internal states of the mind. It is known that in the brain there are no pain sensors; cortex does not feel pain in the same way as, say, a finger when severed. The brain is not a sensory organ and does not have any sensory receptors. How then could emotional states be perceived at all?

According to the theory of William James the quality of emotional states arise from the perception of related bodily sensations. Emotional stimuli are perceived by the sensory areas of the cortex, which then transmit the information to the motor cortex that generates the bodily response [LeDoux 1996 p. 80]. In this theory the feeling follows the bodily responses and is related to the perception of these.

In a different way, the Cannon-Bard theory of emotions proposes that the feeling and the bodily responses occur simultaneously. According to Cannon-Bard an emotional stimulus enters thalamus and is forwarded to the cerebral cortex and to the hypothalamus. The hypothalamus generates the bodily response, while the cerebral cortex generates the feeling [LeDoux 1996 p. 84]. However, this theory does not explain, how exactly the feeling is generated.

Nowadays the pathways of emotional stimuli are known rather well. An emotional stimulus enters the sensory thalamus and is forwarded via low and high routes to the amygdala, which generates the emotional responses. The low route is a direct one, while the high route goes via the cortex. The low route facilitates quick responses to emotionally challenging situations, such as danger [LeDoux 1996 p. 164].

Previously it was stated that only percepts can be conscious. Therefore it seems obvious that also emotional states are perceived via their physiological symptoms, which, in turn are perceived via a large variety of body receptors. Thus the qualia of an emotional state would be the combination of the qualia of the corresponding physical conditions. For instance, anger would be perceived as a state with elevated blood pressure, heart rate, tenseness and possible stomach symptoms, etc. This model of emotional perception is presented in the Fig. 6.1.

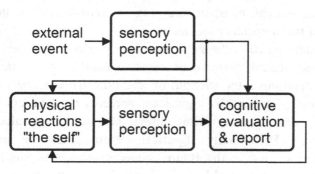

Fig. 6.1. The model of emotional perception. The perception of external events may evoke physical reactions, which are sensed and appear as emotional qualia. Cognitive evaluation may modify these reactions or trigger them alone.

The model of emotional perception of Fig. 6.1 is compatible with the James theory and the Low and High Route model. Sensed external event may excite directly some physical reactions, which are sensed and perceived. The percepts of the external event and the physical reactions are forwarded for cognitive evaluation and the results of this evaluation may modulate the physical reactions. This route allows also the internal evocation of emotional reactions as a result of a thought. The qualia of feelings would be the combined qualia of the physical reactions.

6.3. The System Reactions Theory of Emotions (SRTE)

The author has outlined the so called *system reactions theory of emotions* (SRTE) [Haikonen 2003, 2007], which states that emotions have their roots in elementary sensations. These sensations include good taste and smell, bad taste and smell, pain, pleasure, match, mismatch and novelty. These sensations are characterized by their immediate system reactions. Good taste and smell lead to acceptance, bad taste and smell lead to rejection and withdrawal. Pain leads to withdrawal and possibly aggression. Pleasure leads to sustained action and to the desire of the pleasure-producing activity. Match, mismatch and novelty lead to attention control reactions. More complex emotions arise from the expectations of pleasure (desire, excitement) and pain (fear, excitement) and the fulfillment of the expectations (happiness, disappointment, relief, surprise). In general, emotions are seen as combinations of the system reactions of the elementary sensations.

The system reaction theory of emotions proposes that percepts may have learned neutral, positive or negative *emotional significance* and this value determines the strength of attention that is to be focused on these percepts. Emotionally significant percepts must be attended first and neutral percepts next, if time allows. This emotional value controls also learning. Emotionally significant events must be memorized immediately, so that in the future proper responses to similar events could be generated faster. Memories have also similar emotional values, the so-called emotional soundtrack that accompanies them. This soundtrack is useful when new actions are planned and their consequences are estimated. The good or bad emotional value of the projected outcome helps to guide the planning towards good results.

SRTE is a practical approach that can be readily implemented in systems that have the necessary elementary sensors and utilize associative processing. Some aspects of SRTE have been experimentally implemented by Kawamura *et al.* [2006] in their ISAC robot.

6.4. Emotions and Motivation

Motivation gives us the reason to do whatever we do in a planned manner and in this way motivation powers actions beyond simple stimulus-response reactions. Motivation may arise from internal and external causes. Internal causes arise in response to internal physiological states and needs such as hunger, thirst, high or low temperature, boredom, etc. The common feature of all needs is that the present situation does not correspond to the desired situation. An active internal need evokes the want to strive towards the satisfaction of the need.

On the other hand, reward and punishment are identified as the main external factors of motivation. The expectation of a reward pulls the action towards the given goal, while the fear of a punishment would have the opposite effect. External motivational factors may be taught and learned. In practice internal and external factors may interact in many ways. An internal need may be acute, but cannot be satisfied until an externally imposed task is completed; a hungry worker cannot leave his post at will without fearing a punishment.

Reward and punishment are related to pleasure and pain, which are related to emotions, especially to fear (expectation of pain) and desire (want of pleasure). Emotions are important motivational factors in themselves. We do things out of curiosity, out of interest, fear, anger, envy, jealousy, guilt, revenge etc.

A digital computer does not need motivation. A computer executes exactly whatever commands are programmed, whatever conditional branches are provided. The computer itself does not have a freedom of choice. On the other hand, an artificial associative cognitive agent would not be governed by a program. Many responses would be available at any moment and attention could be directed in numerous ways. No preprogrammed response would exist at any moment, therefore other means of action planning would be necessary. Motivational factors would have an important role in the shaping of the agent's behavior and the agent would seem to have some kind of its own will.

6.5. Free Will

Human consciousness is seen to be accompanied by free will. We are apparently free to think what we like, we are free to make our own choices and decisions. Basically this proposition comes from the dualistic view, which maintains that while the material body has to follow the causal laws of physics, the immaterial mind is free from that kind of restriction and therefore it is free to will whatever it wants. This leads to one argument against conscious machines: Machines can only execute causally controlled deterministic acts. Therefore machines cannot make their own decisions and have free will. Without free will machines cannot be conscious. The obvious weaknesses of this reasoning are the assumptions that 1) free will is a necessary constituent of consciousness and 2) machines cannot execute non-deterministic acts. These assumptions are not necessarily true.

Do we have free will? Free will is the supposed ability to make choices that are free from external influences. In philosophy, an act that is executed by free will is one that does not follow deterministically from the existing conditions, because if it did, then no freedom of choice would have existed in the first place.

We may reject dualism and still maintain that the mind has to have free will, because our everyday experience would seem to prove and demonstrate this. Thus, what kind of experiment could prove that free will exists? This is easy. Tomorrow morning I will prove it by not going to work, even though I would have to. This decision is based on my free will. Quod erat demonstrandum, a philosophical problem is thus solved for good. Or is it? The premises and conclusions of any scientific proof must be scrutinized carefully. Exactly, why did I make this decision? I did it in order to prove the existence of my free will. Thus, my decision has had a cause and obviously only one way of selection; in order to demonstrate my free will, I had to decide not to go to work. If I had decided to go to work as usual, then no point would have been demonstrated, as I had only followed the will of my employer. Alas, my decision was not free, it followed deterministically from the set premises. The existence of free will was not proven.

The author has argued that conscious free will cannot exist [Haikonen 2003 p. 155]. We may buy a certain new car, because its properties suit our purposes and car magazines have recommended it. In fact, we may list our requirements and find out the car that best fits these. This would be our conscious choice. However, this kind of decision making is purely mechanical and could also be executed by a computer. The point is; in conscious decision making we are aware of the ultimate reasons and steps that lead to the decision. However, these steps are causal and consequently the process has been a deterministic one and therefore no free will has been utilized. But, what if we introduced a random factor in the decision making process? Surely this would remove determinism? Indeed, we could flip a coin at one step of the decision making chain in order to break determinism, but would the resulting decision be ours then? Of course not. The decision would be determined by the coin and not by us; the decision would not be a product of our free will.

But, if decisions cannot be based on free will, then what are they based on? Does the negation of free will mean that a cognitive robot cannot make its own decisions at all? Does this mean that we would have to provide the robot with decision-making rules for every possible situation? Do we humans have decision-making rules for every situation? If an exactly humanlike, cognitive conscious robot were only a programmed clockwork, then what would we humans be?

6.6. Decision Making

It can be seen that the decision making process is affected by primary, secondary and external factors. Primary factors arise directly from the situation and include needs, requirements and rational arguments. Secondary factors include emotional state, arousal and alertness, loss of information (forgetting), and own values. External factors include social pressures, the effect of other people, their recommendations and persuasions and any expected reward or punishment. The primary factors are mainly causal and deterministic and therefore do not contribute to the apparent free will. The external factors reflect the will of others and as such do not contribute to the apparent free will, either. Thus the illusion of free will would arise from the secondary factors that are variable and therefore introduce some kind of unpredictability to the process.

Similar factors would contribute to the decision making process of an autonomous cognitive robot. There would not be pre-programmed rules for each decision, because that would be a practical impossibility. Instead, there should be a general value system that could help the robot to choose to do the right thing.

The apparent free will does have a connection with consciousness. The robot should recognize that the decisions are not something that just pop out from somewhere. Instead, the robot should perceive that a situation calls for selections and decisions will and when these are done, they are made by the robot itself. The robot should see that by doing so and by wanting things it is able to alter the course of events. Obviously this kind of self-consciousness would create the illusion of an own free will. This will would not really be free in the strict philosophical sense, but it would be the robot's own will. As a philosophical note, the author argues that the same applies to humans, too. Conscious robots are not programmed clockworks and humans are neither.

Summary

- Emotions are reactions to everyday situations of life and in many cases contain the element of conflict.
- Emotions involve consciously perceived arousal, subjective feeling and physiological symptoms.
- Emotional significance guides attention and learning.
- Emotions motivate; we do things out of curiosity, out of interest, fear, anger, envy, jealousy, guilt, revenge etc.
- Conscious free will does not exist; this does not exclude a subject's own will.

Chapter 7

Inner Speech and Consciousness

7.1. Natural Language

A *natural language* is a symbol system that allows the description of situations by strings of words and sentences. A vocabulary is the collection of the available words and the syntax gives the rules for the indication of those relationships that cannot be conveyed by the basic words alone.

The mastery of a natural language has two parts, namely 1) the ability to understand sentences and 2) the ability to produce meaningful sentences. These abilities are related, but somewhat independent. (It is possible to understand a foreign language without the ability to produce grammatically correct sentences in the same language. Also babies seem to understand speech before they can speak themselves.)

The understanding of natural language text and stories can be demonstrated by asking questions about the subject matter of the text. If these questions cannot be answered, then obviously no understanding has taken place. According to the generative linguistic tradition a language is self-sufficient and requires no references to the outside world; all words can be defined by other words and syntax. Vocabulary and syntax can be formalized and incorporated into a computer program, therefore automated language understanding should be easy. The text to be understood could be stored in the computer memory as a data file. When a question about the text is given, the stored text could be scanned and the requested information could be extracted by word pattern matching and statistical or other algorithms. In simple cases this approach will lead to initial success. However, this approach turns out to be insufficient for large arbitrary texts, because pattern matching and statistical computations do not constitute real understanding. This is definitely not the way that is utilized by the brain, either.

The use of a natural language as a self-sufficient autonomous system will not work. Understanding will not be possible without extensive background knowledge that supplies a contextual framework and the necessary general information that is not included in the story itself. The meaning of words and sentences cannot be defined ultimately by other words, sentences and syntax; this approach would only lead to circular definitions. After all, natural language is used to describe real world situations, therefore the meaning of words and sentences must be grounded to the real world entities. This requirement is apparent in syntactically correct, but ambivalent sentences that can only be interpreted correctly with the knowledge of the related situation.

The view of natural languages as symbol systems with real world grounding of meaning has lead, for instance, to the author's associative multimodal model of language [Haikonen 2003, 2007]. According to this model, the meanings of words are grounded to the percepts of the various sensory modalities and the syntactic structures arise from the relationships in the described real world situations. The multimodal model of language is also an inner situation model.

Inner situation models of language are based on the fact that natural language sentences are descriptions of a real world situations. These descriptions are able to evoke mental visualizations, that is, inner models of the situation. The evoked inner situation models would be similar to ones that are evoked directly by sensory perception, but they may be simpler ones, incorporating only the most salient features.

A perceptual situation model arises from the percepts of the world. Perception produces fleeting percepts that are soon replaced by new ones. However, information about the locations, motions and properties of previously seen, heard or felt objects is retained in short-term memory and this information constitutes the model of the current situation.

In this way, a situation model of environment is created. This situation model is constantly updated and it is also associatively linked to long-term memory, background information and emotional evaluation. The situation model is a model that changes when the situation changes.

Accordingly, Sommerhof [2000] has called these models "running world models".

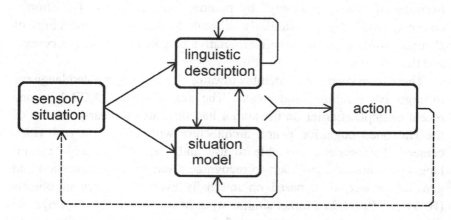

Fig. 7.1. Sensed situation evokes internal interactive linguistic descriptions and imagined situation models. These may evoke action, which may alter the sensed situation. Heard speech evokes directly an internal linguistic description. Beings without language may operate by sole imagination.

According to the situation model of language, inner situation models can also be evoked by verbal descriptions, sentences and stories. The evoked situation model imagery is then processed as if it were produced by direct perception. Verbal questions about the situation direct attention on the relevant details of the evoked model. Therefore it can be seen that words have also the function of focusing attention. Marchetti [2006] has also proposed and developed ideas about the attention focussing function of words.

Zwaan and Radvansky [1998] see situation models as necessary means for language comprehension and they oppose the traditional view of autonomous language; the view that text comprehension could be achieved by the mental construction and retrieval of the text itself alone, without any mental multisensory models of the described situation.

Natural languages have arisen from the need of practical communication, the description of concrete situations. Natural languages are also used to describe abstract concepts, which cannot be explained by simple references to concrete entities; abstract meanings cannot be grounded directly to sensory percepts. It is not possible to explain the meaning of "world economy" by pointing out an object. The simple, common word "but" is similarly difficult to define. The meanings of abstract words arise from their associative connections to larger context and their style of use.

There have been some attempts towards the use of grounded language in cognitive machines and robots. The author's robot XCR-1 is one recent example. Earlier on the author has simulated the early version of the Haikonen cognitive neural architecture with PC and a real video camera. This system was able to use and understand simple natural language sentences and learn to recognize a sentence as a question and give an answer to it, based on internally evoked imagery of objects [Haikonen 1999]. In a rather similar style, Mavridis and Roy [2006] have devised a grounded situation model for robots by using traditional AI means. This model combines language, perception and action in a way that allows the robot to imagine verbally described situations and also to answer questions about perceived and imagined situations.

A natural language also facilitates the "inner speech", which is one's silent self-talk and as such an important facet of thinking.

7.2. Consciousness and Inner Speech

Human mind and consciousness are characterized by a natural language inner speech. Inner speech is the silent self-talk that occurs in our heads when we are conscious. We are aware of our inner speech, which is somewhat similar in appearance to the heard speech. Inner speech is a major component of the contents of human consciousness.

Inner speech is an important form of thinking and is understood as a main cognitive difference between man, animals and machines. In folk psychology inner speech is often equated to thinking even though also non-verbal modes of thinking exist. In inner speech one may comment one's own thoughts and existence and in this way inner speech is related to self-awareness [Morin & Everett 1990, Steels 2003].

The phenomenon of inner speech seems to make us aware of our own thoughts; *if we did not hear our inner speech then how could we know what we think?* Indeed, without inner speech our scope of conscious thoughts would be limited to vague imaginations and feelings.

We are conscious, when we have inner speech, because being conscious is about having internal subjective experience and inner speech is a manifestation of that. This fact might appear as a proof that language and inner speech were necessary preconditions for consciousness. Certain theories of consciousness build on this and propose that especially self-consciousness arises from *thoughts about thoughts*; in our thoughts we can observe that we have thoughts and in this way we become aware of them. These theories lead to the conclusion that animals and babies without a language would not be conscious. However, this is not necessarily so. For instance, to be conscious of pain does not require any special mastery of a language. Moreover, there are also other forms of conscious thinking such as visual and kinesthetic imagination. Thus, a meaningful inner speech may be a strong indication about the possible presence of consciousness, while *the lack of inner speech is not a proof of the lack of consciousness.*

Conscious robots should have natural language inner speech, because in this way their mental processes would be more similar to ours and we could understand better their train of thought. This, in turn, would allow easier human-robot communication.

7.3. Conscious Perception of Inner Speech

In the foregoing it was argued that to be conscious is to have qualia; qualia are the way in which sensory percepts appear internally. There are no percepts without qualia and no qualia without percepts. Thus, the presence and absence of qualia would be the difference between conscious and non-conscious brain activity and consequently it would seem that internal neural activity would have to evoke qualia in order to become consciously perceived.

This leads to the following question: Information processing in the brain is executed by the neurons and their firings. We are not aware of

the firings of neurons as such and obviously in the brain there are no
sensors that would inspect the operation of each neuron. Qualia seem to
be directly related only to the sensory signals, because they depict some
aspect of the sensed entity. Internally generated neural signals do not
have this kind of direct connection to sensors and therefore should be
without qualia. How then can the brain perceive its internal neural
activity in terms of qualia? The same problem occurs also in conscious
robots; how to make them to perceive some inner activity as
imaginations and inner speech, Fig. 7.2.

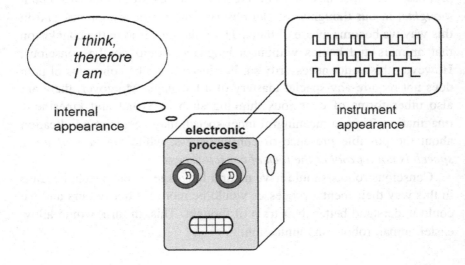

Fig. 7.2. An important step towards true conscious robots: How to make a robot to
perceive some of its internal neural activity as silent "heard" inner speech. The same
problem applies also to inner imagery.

The thoughts that appear internally as the silent inner speech are one
example of consciously perceived neural activity. Obviously in the brain
our verbal thoughts are in the form of neural firings. How can we
possibly "hear" these neural firings as the silent inner speech? This
problem can be solved if we can find a way, in which the internal neural
activity patterns can be translated into qualia based percepts. A simple
trick can do this.

Hearing is based on the use of ears and their inner components that transform the sound vibrations into neural feature signals. These signals appear ultimately to the conscious subject as sound qualia. These neural feature signals have a causal connection to their external sound source; similar sounds produce similar neural feature signal patterns and eventually similar qualia experience, Fig. 7.3.

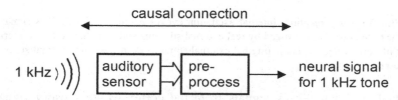

Fig. 7.3. Direct perception of sounds. An auditory sensor transforms the sound vibrations into neural signals. A preprocess performs the frequency analysis and outputs neural feature signals that depict the individual frequencies and their intensities of the sound.

It is known that sound qualia can be evoked also by exciting artificially some of the neural fibers that carry sound information from the ear's cochlea to the cortex. This fact is utilized in the cochlear implants that allow the partial restoration of hearing in some deaf people. It is plausible that also internal excitation of the same neural pathways would lead to the same consequence, the experience of hearing virtual sounds. This idea leads to the feedback loop (sometimes also called re-entry) model of introspection.

Feedback loop models [e.g. Chella 2008, Haikonen 1999, 2003, 2007, Hesslow 2002] propose a feedback loop (or a large number of those) that brings back cortical output signals to the input neurons of the sensory modalities. Here these feedback signals would excite sensory feature signal patterns. The feedback information would be transformed into virtual percepts and in this way the brain could reuse its perception process to introspect its mental content, Fig. 7.4. Thus, if the perception process were able to produce qualia-based conscious experience in the first place, then the mental content mediated by the feedback loop would become consciously perceived in a similar way, too. *In this way one part of the mystery of consciousness would be solved, namely the conscious introspection of mental content.* The feedback loop model can be applied to each sensory modality.

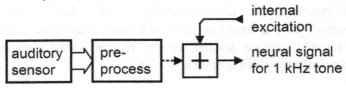

Fig. 7.4. Virtual perception. Internal excitation evokes a sensory signal. This signal is similar to the one that is evoked by real external stimulus and from the system's internal point of view it represents the original external stimulus even when no external stimulus is present.

In Fig. 7.4 feedback signals from the cortex excite sensory signals internally, without the actual presence of the corresponding external stimuli. The excited neural signal patterns are forwarded to the cortex, which will treat them as if they were normal sensory feature signals, produced by real external stimuli, because they occupy the neural pathways from the sensors and the origin of these signals is not inherently encoded in the neural firings. Thus, these feature signal patterns turn into "virtual percepts" of entities that are generated, "imagined" by cortical processes. This principle enables the perception of one's thoughts as the virtual hearing of the inner speech, Fig. 7.5.

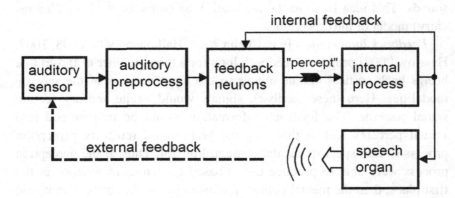

Fig. 7.5. The feedback loop model of the perception of the inner speech. Internal feedback allows the virtual hearing of the inner speech produced by the inner process. Otherwise the inner speech would have to be spoken aloud in order to become perceived.

Figure 7.5 depicts the feedback loop model of the perception of the inner speech. The inner process produces verbal thoughts that are in the form of neural signals that control the speech organ. A thought that is uttered as audible speech by the speech organ can be heard by the system via the external acoustic feedback loop.

Heard sounds are transformed into a set of neural, auditory feature signals by the auditory pre-process. This process includes the spectrum analysis of the sound. The resulting signals are forwarded to the rest of the system through the feedback neuron group. This neuron group passes normally the auditory feature signals from the auditory pre-process, but it also allows the additional excitation of these signals by the internal feedback signals from the inner neuron groups. The excited signals are auditory feature signals and consequently lead to the perception of virtual sounds. However, the virtual sounds differ from real sensory sounds in important respects. Firstly, they are not as vivid as real sounds and secondly, their perception is not associated with any activity in the sensory organs and exploratory sensory acts. The turning of the head will not change the apparent direction of virtual sounds. In fact, there is no associated direction for the inner speech apart from some possible feelings in the speech organ region.

It is possible that in the brain, if this kind of feedback loop existed, the internal feedback would have to be learned in early childhood and before that the only way in which a child could perceive its inner speech would be to talk aloud. At the same time the loop would be rehearsed and learned. (This would explain, why little children must talk aloud all the time, a phenomenon that surprises many new parents.)

The learning could take place in the following way. Let's assume that the inner neuron groups spontaneously generate certain neural signals that cause the speech organ to voice, for instance, "eee". This sound is heard by the auditory sensor and the auditory pre-process outputs a certain set of auditory feature signals, which are forwarded to the feedback neurons. The feedback neurons receive also the internal feedback signals, which coincide with the auditory feature signals and may now become associated with these. Later on, even with the absence of any external sound, the same feedback signals will associatively evoke internally a similar percept of sound "eee". In this way the system may learn to associate a perceived sound with each auditory neural pattern that corresponds to a sound that the sound organ has voiced. Thereafter these imagined sounds may be internally perceived without any overt voicing. The subject will no longer have to talk aloud in order to perceive thoughts in verbal form. (More detailed description in [Haikonen 2007].)

According to the feedback loop model, internal neural firings gain their depiction in terms of qualia when they are fed back to the sensory input neurons, where they evoke the corresponding feature signals. These form the virtual percepts that appear internally in a similar way to the conscious experience that would be evoked by the perception of the real entities.

Summary

- The problem: How to make thought-related neural firings appear as inner speech, similar to heard speech.
- Enabling assumption: The system is able to perceive sensory input neuron firings that are caused by sound stimuli as sounds and sound patterns.
- Solution: Feedback loop models. Reuse the perception process, use feedback loop to excite sensory input neurons internally by thought related inner activity.
- Feedback loop models apply also to vision and other sensory modalities.
- Feedback loops allow the virtual perception of verbal and non-verbal thoughts and in this way solve the mystery of the conscious introspection of mental content.

Chapter 8

Qualia and Machine Consciousness

8.1. Human Consciousness vs. Machine Consciousness

To be conscious in the human way is to have qualia. In the human mind the perceived world presents itself in the form of qualia and so does the internally generated mental content, too. This phenomenal internal appearance is with us, whenever we are conscious.

Consciousness as the presence of a qualia-based internal appearance does not have any executive powers, it is only a special way of internal manifestation of information. Any function that has been attributed to consciousness may be better explained in terms of cognitive functions. These functions can be artificially implemented, at least partly, in various ways without qualia or any other hallmarks of consciousness. Therefore, the mere implementation of cognitive functions is not a sufficient condition for consciousness even if the resulting artifact appeared externally as a conscious agent.

Nevertheless, an information processing method that is characterized by qualia-based internal appearances, may be better suited for the execution of the necessary cognitive functions. Especially, these systems should excel, just like humans, in the direct interaction with the environment aided by memorized experience.

True conscious machines would have qualia-based internal appearances of the externalized world. These machines would perceive the qualities of their environment and their own material body directly, just as we humans do. However, machine qualia would not have to be similar to their human counterparts. In the following some preconditions for machine qualia are discussed.

8.2. Preconditions for Machine Qualia

The first preconditions for machine qualia can be deduced from the fundamental properties of qualia. First, qualia have the ability to appear as apparent world properties instead of appearing as the special neural activity patterns that obviously carry them. The carrying neural mechanism remains transparent to the cognitive entity and only the carried information is perceived in the form of qualia. Secondly, qualia are direct; they are not based on indirect symbolic representations. It is very obvious that systems that do not fulfill these requirements cannot have qualia that are related to the qualities of the sensed entity. Therefore, a perception process with possible qualia should be direct and transparent.

Transparent systems are not unknown in electronics. Phones, radios, television are examples of transparent systems. The user of these systems is not aware of their internal workings.

An audio amplifier is a simple example of a transparent system, see Fig. 8.1. In a sound reproducing system consisting of a microphone, amplifier and a loudspeaker, the input and the output are the same physical phenomenon, namely the sound as the vibrations of air pressure. The audio amplifier system will remain transparent to the listener, if it reproduces the input as such, without adding or removing anything, the only difference between the input and output being the level of sound intensity. The listener will not perceive the internal workings of the system.

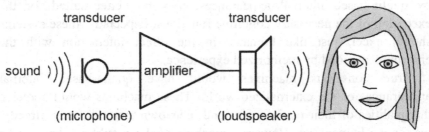

Fig. 8.1. An audio amplifier is an example of a transparent system. The input and the output are the same physical phenomenon and the system remains transparent to the listener if it reproduces the input as such, without adding or removing anything. The listener will not perceive the internal workings of the system.

A *linear transparent system* is of the form:

$$output(t) = k*input(t - delay)$$

where

k = coefficient
t = time point

The output is the delayed version of the input. The delay may be very small ("live broadcast"), but it may also be very long (CD-records, etc.).

The perceiving brain is not a linear transparent system; the sensory output in the brain is not the same physical phenomenon as the sensory input. The eyes receive light, but light is not delivered into the brain. Instead, the light patterns are transformed into corresponding neural activity patterns and the same goes for the other sensory modalities as well. Thus the description of the process would be:

$$output(t) = f(input(t - delay))$$

Thus, the output, the internal neural signal pattern is a function of the input, the sensory stimulus pattern with a very short delay. This kind of a relationship may also be considered as a *transform* or a *mapping* from the sensory stimulus space into the neural signal pattern space. This relationship involves *virtual transparency*; when properly executed, the machinery that executes the transform remains hidden from the output. Obviously, practically every electronic sensor system may execute this kind of a relationship; should we then assume that qualia were present in these systems? Maybe not, but the potential might be there provided that some additional requirements were satisfied.

The author proposes an additional condition and the *amodal feature hypothesis* goes as follows: In perceptual systems qualia may be present if the sensory mapping function conserves the amodal qualities of features. Here the physical input and output phenomena are different, but the patterns of these phenomena are similar.

In the context of the possible electronic realization of qualia-based systems this leads to the question: Do the standard digital data acquisition and signal processing methods conserve amodal features? If yes, then, according to the amodal feature hypothesis, also digital computers with sensors and proper programming could have and experience qualia.

Standard digital data acquisition is based on the analog-to-digital conversion. This process takes temporal samples of the input signal, quantizes them and assigns a binary code word (binary number) to each quantized sample. This code word gives the numeric value for each sample, see Fig. 8.2.

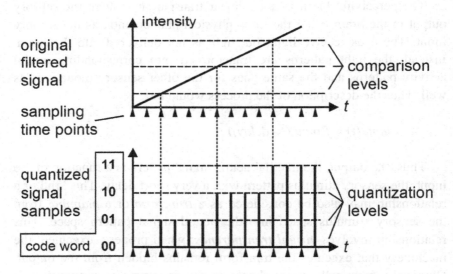

Fig. 8.2. The steps of the analog-to-digital conversion. The input signal is sampled at a given sampling frequency. At the sampling time points the input signal intensity is compared to the fixed comparison levels. The quantized signal level of the sample is determined by the highest comparison level exceeded by the signal. A binary code word is assigned to each quantized sample. In this example the generated digital code sequence is: 00, 00, 01, 01, 01, 01, 10, 10, 10, 10, 11, 11, 11, 11.

The analog-to-digital conversion delivers a sequence of binary numbers that represent the instantaneous value of the digitized signal at each sampling time point. This is a symbolic representation that calls for the additional information about the numeric meaning of the binary code and the information about the represented entity, be it sound, image or some other sensory information. It can be seen that this information is no longer inherently present in the binary data. (For instance, in MPEG digital video coding there are separate code words that indicate the type of the following data packets, be it audio, video or other.) In contrast, in direct transparent or virtually transparent systems this information is carried inherently by the signal and the fixed wiring.

It was earlier argued that symbolic representations do not carry the qualia of the original information and therefore amodal qualia cannot be carried either. Thus, according to the amodal feature hypothesis, digital computers that operate with symbolic representations cannot have and experience qualia. And moreover, computers are not systems that could support qualia-related system reactions.

Audio digital data may be converted back into the original signal by a digital-to-analog converter with arbitrary precision, if the sampling frequency is known and that frequency is at least twice the highest frequency present in the input signal. The number of quantization levels determines the signal-to-quantization noise ratio of the reconstructed audio signal. This allows the reproduction of the original physical phenomenon, such as sound and music, which will evoke the same qualia as the original sound in the mind of a listener, but this fact is beside the original point; the possibility of qualia inside the machine.

The exact nature of qualia may not be completely understood yet, but according to the aforesaid, certain guidelines for machines that may have qualia can be outlined.

1. Qualia are direct, not symbolic representations. Therefore, in order to facilitate qualia, do not make the system perceive via secondary symbols. Secondary symbols may be used in higher stages of cognitive processing.
2. Use direct and virtually transparent perception processes that conserve amodal qualia.
3. Qualia appear to be out there, outside the brain. In order to facilitate this appearance make the system externalize the percepts by letting it to inspect the world via explorative acts.

4. Integrate sensory and motor modalities seamlessly, so that, for example, the direction and location information can be associated directly with the percepts of objects and entities.

The requirements 1. and 2. are the fundamental ones that must be satisfied in any case. The realization of the requirements 3. and 4. without 1. and 2. will not suffice.

Summary

- To be conscious is to experience perception-related neural activity as "internal appearances" with qualia.
- Contemporary computers do not have qualia and are not conscious.
- True conscious machines must have "internal appearances" with qualia.
- Machine qualia may not be similar to human qualia.
- Conscious machines should have perception processes that are direct, transparent and conserve amodal qualia.
- Conscious machines should have perception processes that utilize explorative acts.
- Conscious machines should utilize seamless sensory and sensorimotor integration.

Chapter 9

Testing Consciousness

9.1. Requirements for Consciousness Tests

Basically, there is only one fundamental requirement for real human-like phenomenal consciousness and that is *the requirement of the presence of qualia-based subjective inner experience,* "the internal appearances" of neural or electronic activity. Unfortunately this hallmark of consciousness is available to and is observable only by the subject itself, unless some ingenious methods are invented. Therefore, currently the possible consciousness of a robot can be determined only by indirect ways.

The presence of consciousness in human subjects can be tested quite simply, because we know apriori that the subject is a conscious being. Therefore the subject is either unconscious or conscious (partly or fully) and we only have to find out in which state the subject is at the moment of testing. Thus, for instance, when the subject is apparently asleep or may have lost consciousness due to anesthesia or some medical condition, we may try to arouse the subject and test the following.

- Does the subject have a functioning perception process; does the subject feel pain, does the subject see and/or hear, does the subject respond to stimuli?
- Is the subject able to report anything, verbally or by using bodily signs?
- Are the reports sensible or only reflexes?
- What is the extent of the subject's situation awareness? Does the subject know his/her name, where he/she is, what day it is, etc.

These tests usually work and the subject is more or less conscious if responses are found. However, a subject may fail in these tests and yet be conscious in limited ways. A sleeping person may not respond to these tests, yet the person may be aware of an ongoing dream that she/he may be experiencing. An apparently unconscious patient may be in a *"locked-in state"*, so that she/he cannot respond to stimuli and report pain even though she/he is actually experiencing those.

The detection of artificially generated consciousness is more difficult. *In this case we cannot begin with the assumption that the artifact has a mind with non-conscious and conscious states.* Instead, this is exactly what we seek to find out. A cognitive robot (such as the author's XCR-1 robot) may have various sensory modalities and perception processes and it may generate verbal reports showing that the sensors are acquiring valid information. Yet, without further investigations we may not know how the sensory percepts appear internally; is there any "subjective experience" at all. Therefore the above tests may not be reliable. If a robot fails to respond to these tests then it is more credible than not that the robot is not a conscious agent. On the other hand, if the robot passes these tests (like the XCR-1 robot), it is still not necessarily a conscious agent, because these tests can be passed in many mechanical ways, without any real qualia-based consciousness. Thus, instead of direct tests like the ones above, we have to use indirect methods to determine the presence and scope of consciousness in a cognitive robot.

There are some common sense preconditions for the existence of consciousness in any agent. Therefore the evaluation of a robot's consciousness should begin with the study of the robot's cognitive machinery, the robot's brain. This study should determine, whether the machinery could support the creation and presence of phenomenal qualia-based mental content. Some general criteria exist that do not call for the full understanding of the nature of qualia, see also chapter 8.

General awareness calls for the existence of perception process with a number of sensory modalities. The perception processes must be direct and not be based on symbolic representations, because symbolic representations would exclude the presence of qualia. There is also *the requirement of mental content; an agent cannot be aware of its mental content if there is none.* A conscious agent must also be able to report its mental content to itself; it must be able to introspect in order to become aware of its own thoughts. Perception and introspection call for means of selection, because all information cannot be treated at the same time. Therefore the function of attention must be included. Consciously attended events can be remembered for a while, therefore memory is also necessary.

In addition, a conscious agent should be able to generate responses to the on-going situations and also report its actions and intentions in one way or another to its masters. However, this is a secondary requirement arising from practical causes. An agent or robot without these capacities would not have much practical value. Nevertheless, the ability to report would help in the evaluation of the agent's possible consciousness.

Therefore the cognitive machinery of a potentially conscious agent should support the following general requirements for consciousness:

- Perception, direct non-symbolic
- Mental content
- Introspection
- Attention
- Memory and retrospection
- Responses and reports

The presence of these functions in a cognitive machinery should be rather easy to see. If these functions are not realized within the cognitive machinery of a robot, most probably the robot will not be able support any kind of qualia and be conscious in the human way. However, the presence of these functions does not necessarily guarantee the presence of any consciousness with qualia.

A robot with the above functions, but without qualia and inner experience might behave in a way that would create the external appearance and impression of a conscious agent. Some additional cognitive functions might even lead to the impression of an intelligent agent with free will. This kind of a robot could be said to be *functionally conscious*. However, functional consciousness is not real consciousness, instead it should be seen as some kind of a behavioral simulation of the external appearances of the real thing. Here "functional" means "as if".

9.2. Tests for Consciousness

9.2.1. *The Turing Test*

The Turing test [Turing 1950] is sometimes proposed as a test for machine consciousness, as such or with modifications. In the fifties Alan Turing was considering the question "Can computers think" and if they do, how could we know. In those days computers were readily likened to the brain; they were electronic brains executing mental actions that earlier could only be executed by the thinking human mind. For instance, the conditional IF – THEN program command was naively seen as the computer's ability to make decisions. Turing proposed that a computer could be said to think, if it were able to converse with a test person in such a way that the person would believe that the other party is a human. It should be obvious that this test is badly flawed; to make believe is not the same as to make true. Besides, this kind of fooling tells nothing about the real requirement of consciousness, the presence of qualia-based subjective inner experience.

The so-called *Total Turing Test* [Harnad 1992] is supposed to remedy the obvious flaw in the original Turing test. In order to pass the Total Turing Test the artifact is required to perform exactly as a human being in every empirically testable cognitive challenge; indistinguishable to any judge, and for a lifetime. The totality of this test is supposed to prove that the artifact is indeed a thinking and aware being, cognitively similar to humans. However, the Total Turing Test is based on logical fallacy; it is logically true that an artifact that is cognitively equivalent to humans will pass this test, but it is still logically possible that this test is passed by artifacts that are not conscious and cognitively similar to humans. The Total Turing Test is based on external appearances and as such it does not indicate, which group the tested artifact would belong to. It is of no use to argue that the Total Turing Test would be so exhaustive that no other than those artifacts that are cognitively similar to humans could pass the test, *without also proving first* that the passing of this test by any cognitively non-similar, non-conscious artifact were impossible.

A test that does not work in the first place will not work any better if the scope of the tested property is extended. There is nothing in the

Turing Test or the Total Turing Test that could directly test the presence of any qualia-based inner experience.

9.2.2. *Picture Understanding Test*

According to the Information Integration Theory of Consciousness, proposed by Tononi [2004, 2008], conscious information is integrated and unified. This unification arises from a multitude of interactions between the different parts of the brain. If the interactions cease, as during anesthesia or deep sleep, consciousness disappears. According to Tononi, a subject needs a large repertoire of actively connected information in order to be conscious. Based on that, Koch and Tononi [2011a, 2011b] have proposed that this condition can be used to test consciousness in machines by letting them to try to understand pictures.

Koch and Tononi propose an example: Let us suppose that there were two test pictures. One picture depicts a computer display with a keyboard in front of it, the other picture depicts the same computer display with a flower pot in front of it. According to Koch and Tononi, a non-conscious computer with machine vision would not realize that there is something wrong with the second picture, because the computer would not have and would not be able to evoke the necessary background information about the relationships of the items in the picture. Therefore the computer would not really understand what it sees. On the other hand, we, conscious humans, would immediately see and recognize the untypical situation due to our vast amount of background information and our capacity to integrate it meaningfully.

It should be obvious that this test is the Turing test in disguise. To pass this test the machine should exhibit similar cognitive abilities in picture understanding as humans. However, there nothing in this test that could prove that the machine was actually conscious; this test addresses only certain issues of computer vision algorithms and the integration of any available background information. The example test could be easily passed by artificial neural classification, by storing a large number of pictures of computer displays with keyboards. Thereafter a picture of a computer with a flower pot would surely stand out as an exception. No consciousness would be involved here.

As such, this test does not have much to do with consciousness, because the essential issues of qualia and inner experience are ignored.

9.2.3. *The Cross-Examination Test*

It seems that currently there is only one way of finding out the presence of qualia-based inner experience in a robot; we must ask and cross-examine the robot itself. Unfortunately this method would exclude those robots and machines that do not use and understand a natural language; this is a similar problem to that when we try to find out whether animals have any inner experiences.

If the robot is able to report that it has the flow of inner imagery and inner speech and we know that this report is not a pre-programmed one and the internal architecture of the robot's brain could be able to support these, then we should determine that the robot is actually able to introspect its mental content and is able to report it; the robot is conscious. If the robot is also able to report that it has inner speech, which appears as a kind of heard speech, then we should determine that the robot has some kinds of internal appearances and qualia. The cross-examination test may focus on the following issues:

- Does the machine have a flow of mental content that is about something?
- Is the machine able to report its mental content (percepts, thoughts, inner speech etc.) to itself (and others) and does it recognize the ownership of the same?
- Is the machine able to describe some qualia?
- Does the machine remember its immediate past?
- The "hammer test" of phenomenal awareness: Does the machine feel pain? In which way?

One may wonder, if the cross-examination test were just another variant of the Turing test. In the Turing test the actual realization of the tested machinery is not considered. The cross-examination test, however, is meaningful only for architectures that are known to fulfill the basic requirements for potentially conscious machines.

9.3. Tests for Self-Consciousness

9.3.1. *Self-Consciousness*

Self-consciousness is the subject's ability to recognize its own existence as a discrete being. A self-conscious subject is aware of its own existence and has some kind of a *body image* and a *mental self-image*. Self-consciousness includes also a concept of *first person ownership*; I am aware that my body is mine, my thoughts are mine, my memories are mine, my decisions are mine, my speech is mine, my acts are mine.

The body image relates to the shape, size and function of the body and the difference between the body and the environment. In humans and animals the border between self and the environment is usually clear. The body belongs to the subject and the environment does not. The body is neurally wired to the brain, while the environment is not.

The neurally wired network of various body sensors is called the somatosensory system. This system consists of several sensory modalities with a large number of receptors that sense body part position (proprioception), touch (tactition), pain (nociception) and temperature. The somatosensory system allows the brain to monitor the status of the body continuously. The information delivered by the somatosensory system leads to some simple rules: If it hurts, it is a part of mine, not to be eaten by me. If I can feel it, it is mine. The first rule is seemingly rather ridiculous, but it is not, as is witnessed by the unfortunate sufferers of the congenital insensitivity to pain (CIPA). These subjects do not feel pain and may actually try to eat their tongues and lips, etc. The second rule is manifested, for instance, during an occasional numbness (paresthesia) of a hand due to a bad sleeping position. It may then happen that for a scary moment the subject recognizes the hand as a foreign object.

The mental self-image relates to the ideas that the subject has about itself; who am I, how do I feel, what do I want, how am I in respect to others. The mental self-image calls for the retaining of personal history in the form of memories.

The group of ideas, which contains a body image and a mental self-image that are grounded to the material self via a somatosensory system, can be called the subject's *self-concept*. A self-conscious subject can refer to itself in many ways within the self-concept group of ideas.

Based on the foregoing the prerequisites for self-consciousness can be listed. In addition to the general requirements for consciousness a self-conscious agent shall have:

- Somatosensory system
- Body image
- Mental self-image
- Remembered personal history

The testing of self-consciousness in a robot should start with the verification of the presence of the equivalent of the somatosensory system and memory systems with the functions of introspection and retrospection. This can be verified by the inspection of the cognitive architecture of the robot. If these are implemented in the robot then the presence and extent of a body image and a mental self-image should be determined. Sometimes the block diagram of a cognitive system may contain "boxes" with the labels "body image" and "self-image". These boxes should not be taken at the face value. The presence of the body image and mental self-image should be tested by functional means, for instance, by checking the robot's ability to refer to its body and self-related mental content.

9.3.2. *The Mirror Test*

The so called *mirror test* is sometimes used to test the self-consciousness of babies and recently also animals [Gallup 1970]. The test subject is placed in front of a mirror and the subject's behavior is observed. The test is passed, if the subject recognizes itself in the mirror. Many animals and very young babies seem to interpret their own image in the mirror as another being and may try to look behind the mirror. In these cases the test is clearly not passed. Sometimes, however, it may be difficult to determine directly, whether the subject recognizes itself in the mirror and therefore variations of this test may be used. In the *rouge test* a red dot is secretly placed on the forehead of the subject and then a mirror is placed in front of the subject. The rouge test is passed if the subject realizes that the red dot on the mirror image depicts the red dot on the subject's face. Human babies of the age of 18 months or over usually pass the mirror test. This is also the age, when babies usually become socially self-aware [Lewis *et al.* 1989]. Some animals including chimpanzees, elephants and possibly pigeons can also pass the test. In strict sense, the mirror test only indicates that in some cases the test subject is able to recognize itself in the mirror image. This is a cognitive task that may fail even when the subject possessed some kind of self-concept. The mirror test has been applied to robots by, for example, Takeno *et al.* [2005].

9.3.3. *The Name Test*

The *name test* is based on the assumption that if a subject has a self-concept, then this may be associated with a name. Little children learn their names early and may use those instead of "I" to refer to themselves in their speech. Many animals seem to be able to learn their given names, but it is not clear, whether they associate this name with themselves or with an event. A cat may come when called by name, but this may happen, because in the cat's mind the name is actually associated with food or petting and not with the cat itself. Animals without the ability to speak cannot use their names in overt speech and thus we will not know, what kinds of associations have taken place. Therefore, the passing of the name test is not necessarily a reliable indication of self-consciousness.

The name test can be used with robots, too, if the auditory modality allows the recognition of at least some spoken words. If an advanced robot has the faculty of a natural language then it should be possible to observe, if the robot learns its name and uses it to refer to itself. But then, it would also be possible to ask the robot about its concepts of self and in this way the issue of the robot's self-consciousness would be resolved.

9.3.4. *The Ownership Test*

A self-conscious subject should posses the concept of the ownership of the body, thoughts, memories, decisions, speech, acts and possible acquired possessions. If the subject is not able to communicate with a natural language, then the presence of the concept of ownership must be determined by indirect means. Does the subject recognize that a certain act has been executed by the subject itself? The determination of this may not necessarily be difficult. Consider, for instance, a dog that has destroyed few household items just for fun while left alone home. When confronted by the master, the dog readily shows the apparent signs of guilt and shame; the dog knows, who has done all this.

9.3.5. *The Cross-Examination Test*

The cross-examination test may also be used to determine the scope of the self-consciousness in a robot that is able to communicate verbally. For instance, the following issues may be considered:

- Is the machine able to make the difference between the environment and the machine self?
- Is the machine aware of its own existence? In which way?
- Does the machine bind its present situation to the personal history and to the expected future?
- Self-description; body image and mental self-image. How does the machine perceive itself?
- Ownership; is the machine able to declare any ownership?

9.4. Requirements and Tests for Machine Consciousness in Literature

9.4.1. *Aleksander's Axioms*

Aleksander and Dunmall [2003] propose five axioms that would define a minimal set of material preconditions for a conscious biological or non-biological agent. These axioms are:

1. Depiction. The agent must have internal perceptual states that depict external world and the agent itself.

2. Imagination. The agent must have internal states that depict recalled or imagined depictions of the style of the axiom 1.
3. Attention. The agent must be able to choose which parts of the world or the agent itself is internally depicted at a given time.
4. Planning. The agent must be able to imagine future action and select among imagined alternatives.
5. Emotion. The agent must have affective states that evaluate planned actions and determine the selection of the action to be executed.

Obviously humans usually satisfy these requirements and are also conscious. The first requirement, depiction, would seem to be the most important one, because without internal perceptive depiction no cognition could take place. However, humans with limited faculties of imagination, attention, planning and emotion would still be counted as conscious persons albeit not very bright ones. On the other hand, a cognitive machine might seem to posses these capacities and yet be without any real consciousness; it would all depend on the nature and style of the internal depiction. Thus Aleksander's axioms may be considered as a useful list of preconditions for consciousness, but nevertheless, the fulfillment of these preconditions in a system does not guarantee the presence of consciousness in that system.

9.4.2. *The ConsScale*

Arrabales, Ledezma and Sanchis [2010] propose that consciousness is a rather continuous phenomenon, starting from a very weak and dim minimal consciousness and extending to a possible super-conscious state. Accordingly, Arrabales *et al.* have devised a biologically inspired scale for the assessment of the level of the functional consciousness in machine consciousness implementations. This scale, *ConsScale*, is a partially ordered set of requirements that is based on a particular dependency hierarchy for cognitive skills.

ConsScale defines the following levels that are considered relevant to the functional aspects of consciousness: 2 – Reactive, 3 – Adaptive, 4 – Attentional, 5 – Executive, 6 – Emotional, 7 – Self-conscious,

8 – Empathic, 9 – Social, 10 – Human-like, 11 – Super-conscious. For each level a number of abilities are listed. Based on these ConsScale offers a rating system for the formal evaluation of the functional consciousness in a given cognitive machine. However, the ConsScale method does not address the phenomenal appearance of consciousness in any direct way. It is obvious that the extent of consciousness, which is known to exist (like in humans), can be evaluated by ConsScale. However, the evaluation of artificial agents with ConsScale tells only something about the extent of cognitive capacities of the evaluated agent. The passing of the ConsScale tests does not prove that the artificial agent were phenomenally conscious to any degree.

Summary

- We are conscious; therefore other people are also conscious beings because they are biologically similar to us.
- Robots are not biologically similar to us; therefore we cannot assume that robots, which may behave like humans, would be similarly conscious.
- The primary target for a robot consciousness test would be the presence or absence of qualia-based inner experience.
- If the robot to be tested is able to converse verbally, then the "cross-examination test" may be used. However, we must know that the robot has an internal architecture that may allow qualia and consciousness.
- Tests for self-consciousness should focus on the "self-concept", "body image" and "mental self-image".
- Tests for self-consciousness include the "mirror test", the "name test" and the "ownership test".

Chapter 10

Artificial Conscious Cognition

10.1. Which Model for Artificial Cognition?

In the course of time there have been various proposals for the technical model of the brain and the mind.

1. *The brain is a switchboard.* One of the earliest technical models was that of a telephone exchange [Pearson 1911/2007]. Neural signals were known to exist and the brain was seen as a switchboard that connected sensory, motor and other nervous resources together according to the requirements of each situation.

2. *The brain is a computer.* In the fifties the emerging electronic computer was seen as a suitable model for cognition. After all, the computer executed tasks that could be executed earlier only by the human mind. Thus, the brain is the hardware, the mind is the software and cognition is symbolic computation, the execution of algorithms.

3. *The brain is a hierarchical classifier.* Early artificial neural networks were classifiers and pattern recognizers and quite conveniently cognition and memories were seen to be based on hierarchical classification and categorization [e.g. Estes 1994].

4. *The brain is a controller.* The brain controls our body and behavior. Feedback control systems are well-known in engineering. Therefore, the brain might be modeled in terms of advanced predictive model-based feedback control systems. Complex hierarchical feedback systems may even be seen to give rise to self-awareness and consciousness; for instance, Sanz has proposed that a model-based cognitive feedback control system is conscious if it is continuously

generating meanings from continuously updated self-models [Sanz, López and Bermejo-Alonso 2007, Sanz *et al.* 2007].

5. *The brain is a predictor.* The ability to anticipate and predict what may happen next gives a survival edge to any being. For instance Edelman [2008] has proposed that mind is a tool that allows the prediction of the world and this in turn allows meaningful planning of actions.

6. *The brain is a simulator.* The real world is outside and the mind is inside the head. The contents of the mind are necessarily only reflections, simulations or models of the real world. For example, Hesslow [2002] has proposed that conscious thought is a simulation of motor acts and perception. Mental models have been seen as necessary preconditions to consciousness [e.g. Holland and Goodman 2003, Holland, Knight and Newcombe 2007].

7. *The brain is a search engine.* At each moment the brain is looking for a suitable response to the current situation. The continuation of each verbal thought calls for the searching for the next word. The meanings of seen objects pop out instantly without any perceived effort. These tasks are typical search tasks and the brain is executing them in a most efficient way.

8. *The brain is a time machine.* The brain is able to execute mental time travel (chronesthesia). This is the mind's ability to think not only about the present, but also the past and future. This ability is closely connected to situation awareness and self-consciousness. Patients with hippocampal damage are not able remember past episodes or imagine future. These patients live in constant present moment with a time span of the order of only tens of minutes [see e.g. Suddendorf, Addis and Corballis 2009].

All these models have their merits, but which model is the correct one? The answer is, of course, that they all are, but only to a certain extent and not alone. A complete model for artificial cognition would have to combine the pertinent features of the above models into one integrated framework.

However, there is an additional essential issue that applies to all models of artificial cognition. This is the question of sub-symbolic and symbolic information processing.

10.2. Sub-symbolic vs. Symbolic Information Processing

The brain is a biological neural network. Every researcher, who is familiar with artificial neural networks, knows that you cannot run programs on neural networks. Therefore the brain cannot be a computer. On the other hand, every computer programmer knows that every computer program can, at least in principle, be executed mentally, perhaps with the aid of pen and paper. One can read the program code and mentally execute each command exactly as the computer would do. If this were impossible, then computer programming would also be impossible. So, the brain can run programs after all, even though it is a neural network. This leads to the questions: What is the difference between the biological neural network brain that can run programs and the artificial neural networks that cannot run programs? What is the ingredient that the brain has and the artificial neural networks do not have? The simple and obvious answer can be found by inspecting the difference between computers and artificial neural networks.

According to the Artificial Intelligence paradigm, cognition and possibly also consciousness are computational. The mind is based on symbolic computation, not unlike the computations executed by a digital computer. This proposition follows from the hypothesis that only symbolic computation can produce results similar to those produced by human cognition. This is the only way and "there is no other game in the town" [Fodor 1975].

On the other hand, according to the connectionist neural network approach, cognition and consciousness are based on sub-symbolic processing, which cannot be executed by a digital computer. After all, the brain is a neural network. However, this does not exclude the possibility to simulate sub-symbolic neural systems by computer programs. It goes without saying that so far, both the Artificial Intelligence and the traditional artificial neural network approaches have been remarkably unsuccessful in their quest for general intelligence and machine consciousness. What has been missing, will become obvious later on.

Computers process *symbolic representations*, while traditional artificial neural networks process *sub-symbolic representations*. The difference between these representation methods can be illustrated by an example. Consider the following visual patterns:

T A L O

In a visually perceiving sub-symbolic neural system the visual patterns *T, A, L, O* will activate certain neural signal patterns. These signal patterns will be present, when the corresponding visual pattern stimuli are present. These signal patterns are called sub-symbolic, because they "represent" the corresponding visual patterns directly, without any additional associated information. The meaning of a sub-symbolic sensory signal is grounded to the source of the signal, such as a feature detector of a sensor. Other examples of sub-symbolic representations are qualia including pain. Qualia are direct and experienced without any further interpretation. Pain is pain and it hurts without any additional explanatory information.

In a symbolic system, the visual patterns *T, A, L, O* may be taken to be symbolic representations that represent the letters *t, a, l, o*. The combination of these letters may be taken to represent a word. These meanings cannot be found by inspecting the details of these visual patterns, they must be otherwise known. In this case, no matter how you try, you will not understand the meaning of this word unless you know that it is the Finnish word for "a house". Thus, the understanding and interpretation of a symbol calls for additional information that must be available to the interpreter. A symbol may depict various things, such as objects, entities, commands and rules. The meaning of a symbol is fixed by convention, not by any direct grounding to sensory percepts or the like. The ability to use symbolic representations is *a necessary requirement for natural languages*, but it is also more.

The ability to utilize symbols with attached rules is the primary requirement for algorithmic computation. A computer program is an algorithm, a sequence of instructions, which leads to the desired outcome, when executed exactly in the specified way. In digital computers, the symbols are in the form of binary words and may represent machine code or data. Code words make the computing machinery execute the required operations on the data. Different programs make the same machine execute different tasks, therefore *the computer is a versatile multipurpose machine.*

Traditional artificial neural networks do not operate with symbols with attached rules and therefore cannot run programs. The only algorithms that they apparently execute, are those for the adjustment of their synaptic weights during learning and those for the generation of outputs, when certain inputs are given. These algorithms operate in the sub-symbolic domain and are fixed. The well-known Back-Propagation algorithm and various self-organizing algorithms are examples of learning algorithms for artificial neural networks. These algorithms allow the execution of one task within the given network, for instance, the classification of input patterns, but nothing more. *Traditional artificial neural networks are single purpose circuits.*

In order to be able to execute computer programs, the brain has to be able to operate with symbols. On the other hand the brain is a biological neural network that is able to operate with sub-symbolic representations such as qualia, including pain. Technically, sub-symbolic representations are simpler. Based on that, one may assume that in the brain the sub-symbolic way is the primary way of operation, but there would also be a mechanism or effect that would allow the transition from sub-symbolic representations to symbolic representations. Traditional artificial neural networks do not have this kind of a mechanism. Therefore they cannot utilize symbolic representations and execute algorithms that are controlled by symbols with attached rules.

In the brain, information is represented by neural signals. It is obvious that the signals from sensors are sub-symbolic, but what kind of representation would be used for symbols? The author proposes that in the brain sub-symbolic and symbolic representations are not different, they are similar neural signal patterns. Only their usage is different. Those neural signal patterns that are used as symbols, have associated meanings. Consequently, *the mechanism that facilitates the transition from sub-symbolic representations to symbolic representations is association.*

This conclusion leads to *the rejection of both the computer and the traditional artificial neural networks* as potential platforms for artificial conscious machines. Instead of those, this conclusion calls for a different approach, namely that of *associative symbolic neural networks* that are used with a special *cognitive architecture*.

10.3. What Is a Cognitive Architecture?

Cognitive architectures depict system arrangements for the production of human-like cognition. Cognitive architectures try to combine and implement a large number of cognitive functions within a single system and may be considered as functional models for the biological brain or as blueprints for robot brains. Typical *cognitive functions* that may be included are:

- Perception
- Attention
- Prediction
- Learning
- Memory
- Imagination
- Planning
- Judgment
- Reasoning
- General intelligence
- Emotions
- Natural language
- Motor action control

Early Artificial Intelligence approaches implemented only one or few aspects of intelligence and consequently were quite fragile; the approach failed outside the narrow scope of its competence. The targeted *general intelligence* was not, and has still not been achieved. *Cognitive approaches* try to remedy this by integrating many cognitive functions in a system and in this way try to achieve a more universal cognitive competence, which would be comparable to that of humans.

An advanced cognitive architecture should also address the issues of *consciousness, self-consciousness, subjective experience* and *qualia*. A robot with an implemented cognitive architecture should produce human-like responses, action and behavior. This robot should know and understand what it is doing and it should be able to act autonomously in meaningful ways in new situations. It should also be able to learn and accumulate experience and in this way become more proficient and skillful during the course of time.

Cognitive architectures may be based on different approaches.

The high-level symbolic approach defines the action of the implemented cognitive functions directly and may assign a dedicated system block for each cognitive function. These function blocks are then logically connected to each other. This approach is usually implemented as a computer program and represents typical AI tradition.

The low-level sub-symbolic approach begins with the definition of the function of fundamental components such as the neuron and synapse and their artificial implementation. Next, the functions of interconnected neurons and neuron groups are defined. The desired high-level cognitive functions should eventually somehow emerge as the result of these interconnections. This is the traditional artificial neural network approach, known also as *the connectionist approach.*

The associative symbolic neural networks approach begins also with the definition of the function of the fundamental components such as the neuron and synapse and their artificial implementation. However, in this case the neurons differ from those of the traditional artificial neural network approach. The neurons and neuron groups used here are associative and they allow the association of signals and signal patterns with each other in a direct way. This approach allows the association of meaning and the transition from sub-symbolic to symbolic level in a natural way.

A cognitive architecture may include a master executive and/or a master attention control. In other cognitive architectures these operations may be distributed along the system.

Cognitive architectures are often described with the aid of *block diagrams*. In engineering, block diagrams are very useful, when a general view of a complicated system must be presented. The blocks of a block diagram represent functional modules and the connecting lines between the blocks represent the information flow and functional connections between the modules. A block diagram is a valid presentation only if the inputs, outputs and the executed functions inside the blocks are defined. In software systems a block represents a program routine and in electronic hardware realizations a block represents a circuit diagram. A circuit diagram (schematic diagram) depicts the actual connection of electronic components such as transistors, resistors and capacitors for the execution of the desired function. A block may have several alternative ways of implementation. A block diagram without well-defined blocks and interconnections has only entertainment value.

A large number of cognitive architectures have been developed and proposed during the last decades. For instance, Samsonovich [2010] lists 26 architectures in a catalog of implemented cognitive architectures and even this list is not complete. Duch *et al.* [2008] give a compact review of the different approaches of cognitive architectures.

In the following the associative symbolic neural networks approach is discussed. The described approach utilizes *associative information processing* with special artificial associative neurons and a special cognitive architecture, the *Haikonen cognitive architecture*.

Summary

- Various models for artificial cognition exist, including feedback control systems, the computer and artificial neural networks. All these have merits and shortcomings.
- Conscious cognition calls for sub-symbolic and symbolic processing. This leads to the rejection of both the computer and the traditional artificial neural networks as the probable platforms for artificial cognitive and conscious machines.
- Associative symbolic neural networks with cognitive architectures and associative information processing are proposed as the solution.

Chapter 11

Associative Information Processing

11.1. What Is Associative Information Processing?

What is thinking? Introspection seems to reveal that our thoughts are not a random stream of separate reflections. Instead of that, each new thought appears to be linked to the previous one by some connection; thoughts are associated with each other. This phenomenon was noted by Aristotle, already some 2500 years ago. Aristotle also tried to find the basic principles of association in human thinking. (A summary of Aristotle's ideas is presented by Anderson [1995].)

Much later on, David Hartley (1705–1757) and David Hume (1711–1776) developed the philosophical theory of *associationism* further. They proposed that independent ideas in the form of sensations and mental images were associated with each other via contiguity and resemblance. For instance, a present sensory percept may evoke memories of similar occurrences in the past. Simple ideas were considered as kinds of atoms of thoughts. During thinking, the combination of these atoms by the rules of association would give rise to the flow of thoughts. In short, the theory of associationism proposed that association was the basic mechanism of mental activity.

Later on the shortcomings of the theory of associationism have been pointed out, for instance, by Henri Bergson in his book *Matière et mémoire* [Bergson 1896/1988]. Bergson argued that in a large associative system almost everything may be eventually associated with everything enabling an infinite number or possible recollections that resemble in some way or other the present set of percepts, yet only one of them should consciously emerge. How would this selection possibly take

99

place? According to Bergson, associationism cannot explain this or provide a sufficient selection mechanism. Contiguity and resemblance may have a role in thinking, but they are not sufficient mechanisms for cognition.

The *associative information processing method* that is proposed here is designed to facilitate the transition from sub-symbolic to symbolic processing in artificial neural networks. This approach is different from the trivial associationism. This method is based on associative connections, but is advanced beyond the trivial associationism by having selective mechanisms that control associative evocations. These mechanisms rely on the strength of evocation, context and emotional significance and they operate via various and adaptable thresholds. Here the associative mechanisms are to be found in the neural machinery and they learn their associative neural signal connections via coincidences and correlations. The high-level results of this process may sometimes appear as if they were caused by contiguity and resemblance; many other times such impression is not present. This associative information processing method is not to be confused with the traditional artificial neural network methods that require hundreds or thousands of learning examples for proper training.

This associative information processing method operates with distributed neural signal representations, is able to learn correlations, utilizes associative linking and evocation, facilitates associative memories and enables naturally the transition from sub-symbolic to symbolic processing. It is proposed that this kind of associative information processing is sufficient for cognition, when used in a proper cognitive system architecture with sensory and motor modalities.

11.2. Basic Associative Processes

11.2.1. *Pavlovian Conditioning*

Simple association connects two things with each other, so that one may later on evoke the other one. Early in the last century Russian physiologist I. P. Pavlov pioneered research on simple stimulus-response association in animals, which was later on known as the *Pavlovian conditioning* [Pavlov 1927/1960]. Pavlov's experiments can be seen as a proof of simple associative processing in the nature. The principle of Pavlovian conditioning is depicted in Fig. 11.1.

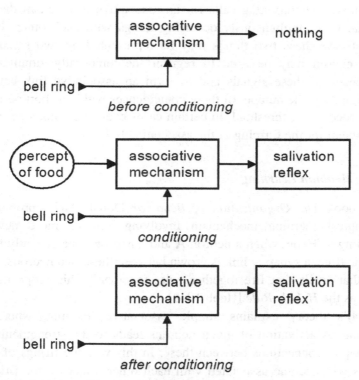

Fig. 11.1. Pavlovian conditioning with an associative mechanism. Bell ring and food are associated with each other in a dog's mind. Before conditioning, a bell ring will not evoke salivation. During conditioning the bell ring is associated with the percept of food. After conditioning, the bell ring alone will evoke salivation as if food were present.

Pavlovian conditioning is a simple case of associative learning and associative evocation. In Pavlov's original experiments with dogs, food and the ringing of a bell were associated with each other in a dog's mind by repeated simultaneous occurrences. Afterwards Pavlov noted that the conditioned dog would begin to salivate whenever he rang the bell even when no food was present. The ringing of the bell had become a *conditioned stimulus* for the salivation response. Instead of the ringing of a bell, many other stimuli may be turned into conditioned stimuli by repetition. (According to some reports, at older age Pavlov had an unexplainable urge to ring a bell, whenever he saw a salivating dog.)

Pavlovian conditioning calls for an associative mechanism that can associate two simultaneously occurring things with each other. In the simplest case these two things may be represented by two signals. A simple neuron may be used to register the temporally simultaneous occurrences of these signals and to form an associative link between these signals, if the number of these coincidences within a short period of time exceeds a set threshold. In certain cases even one coincidence may be sufficient for the forming of the associative link.

11.2.2. *Hebbian Learning*

In his book *The Organization of Behavior* Donald Hebb proposed a neurological learning mechanism involving two or more neurons. According to Hebb, when a neuron A and a neuron B persistently fire at the same time, a synaptic link is grown between these two neurons. Hebb noted that "cells that fire together, wire together." This proposition is known as the *Hebb's Rule* [Hebb 1949].

Hebb's theory explains simple associative learning, where the simultaneous activation of given neurons leads to the strengthening of the synaptic connections between these. In this way the firings of these neurons will become associated with each other. Later on the firing of one neuron will lead to the firing of the other neuron, too, due to the learned synaptic connection and in this way the original neural activity pattern is recreated.

Hebbian learning can be seen as a simple neurological explanation for Pavlovian conditioning in the context of two or more neurons. The principle of Hebbian learning can also be mathematically formulated in different ways. Various Hebbian learning algorithms for the determination of synaptic connection strengths in artificial neural networks exist [Jain *et al.* 1996]. Hebbian learning is also used in associative memories.

11.2.3. *Autoassociation and Heteroassociation*

Associative processing is based on the association of an item with another, which then allows the evocation (recall) of the other, when the other is given. There are different forms of associative links.

In *autoassociation* the parts of a pattern are associated with themselves. Thereafter a part of the pattern will evoke the whole pattern. A network that executes autoassociation is called the autoassociative memory. Autoassociative evocation facilitates the imagination of the hidden parts of visually seen objects, Fig. 11.2.

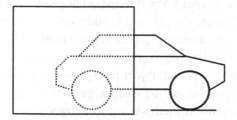

Fig. 11.2. Autoassociative evocation facilitates the imagination of the hidden parts of an object.

Heteroassociation refers to the association of unrelated patterns with each other. Examples of heteroassociation are Pavlovian conditioning, association of meaning and the use of symbols.

Temporal association relates to temporal sequences of patterns. In *temporal autoassociation* a number of temporally previous patterns are associated with the following patterns and thereafter the previous patterns will be able to evoke the following patterns. Temporal autoassociation allows the prediction and recall of the rest of a temporal sequence, when an early part of it is presented as a cue. Temporal autoassociation is used, for instance, in the memorization and remembering songs by heart. In *temporal heteroassociation* a pattern or a sequence is associated with another sequence, so that the presence of this pattern or a sequence can evoke the associated sequence. Temporal autoassociation and heteroassociation may work together.

All these forms of association are needed in associative processing.

11.3. The Representation of Information

What would be a suitable information representation method for associative systems? There are several different information representation methods. In digital computers information is represented by binary numbers. Images are represented by pixel (picture element) maps, where each pixel is a binary number that indicates the luminous intensity of the picture at that point. Sound is represented by a series of binary numbers that indicate the intensity of the sound at each time point. These methods of representation have their virtues, but they do not readily describe the content. Pixel maps do not tell what kinds of objects might be in the image, digital audio code does not tell what kinds of sounds, speech or other, etc. might be represented by the code. In digital signal processing the issue of *content detection* is a difficult one that calls for special techniques. Visual object recognition relies on various pattern recognition methods that try to extract pattern information from pixel maps. Speech recognition methods try to extract phoneme patterns from the stream of binary numbers that constitute the audio code.

It is obvious that a cognitive system must be able to execute successful content detection, but even this is not sufficient. The extraction and detection of visual or auditory patterns are only the first step towards cognition. The next step would involve the *detection of meaning*. The meaning is rarely present in the detected patterns. Instead, it is something that is associated with the pattern in the mind of the observer. Only learned experience can give meaning to the perceived objects and sounds. Printed words remain only meaningless ink stains on the paper unless you have learned to associate the ink patterns with letters, the combinations of letters with words and the words with the entities that they represent. This is a task for associative processing.

Due to the problems with content detection and the detection of meaning, the traditional digital representation of information with binary numbers is not well suited for cognitive systems. A better representation method would be one that would directly tell something about the content and would allow easy association of meaning. This method should also allow easy modification and combination of these representations so that imagination and intelligence would be possible.

This method should also work with imperfect representations and should tolerate errors and distortion. These kinds of representations exist and they are called *distributed representations*. Pioneering work on distributed representations has been done by Hinton, McClelland and Rumelhart [1986]. Also Kanerva has done related groundbreaking work on associative sparse distributed memories [Kanerva 1988].

11.4. Distributed Signal Representations

Each sensed object has a variety of features like the shape, size, color, surface texture and so on. The combination of features give the object its specific appearance and therefore objects can be described and identified by listing their observable features. A feature may be described by sub-features, but eventually this division into sub-features leads to a feature that cannot be described as a combination of other features as it has only one indivisible property. This kind of a feature is called here an *elementary feature*. An elementary feature is property that can be represented by one on/off signal; the so-called *feature signal*. When the property is present, the signal is on, otherwise the signal is off. In a perception process elementary features are detected by specific elementary feature detectors that output feature signals when ever they detect their specific elementary feature.

A feature signal may be binary having these two values only, but it may also be continuous. In that case the intensity of the signal may be used to convey the confidence value of the observation or the importance of the observation.

Any object can be described by a collection of elementary features and consequently, by an array of feature signals. These signals tell "what is where" and "what is when". A cherry can be described as a small, red and round object with a sweet taste. However, there may be also other sweet small, red an round objects, like candy balls or even candy replicas of cherries. Exact recognition is a tricky thing, but luckily in everyday situations context usually provides the necessary additional information.

Distributed signal representations can be used to depict any property that can be detected by a suitable detector. Such detections may be visual, auditory, tactile, etc. and they may indicate quality, position, motion, relative sizes, changes of quality, position, motion and so on.

A distributed signal representation is descriptive. It tells directly what is represented in terms of elementary features and their combinations. As such it is direct and if realized in a virtually transparent way that conserves amodal features, it may also fulfill the proposed requirements for machine qualia, see chapter 8.

Distributed signal representations allow easy modification of the appearance of the depicted objects and action. Each property can be changed independently and completely new objects can be depicted by combining feature signals in novel ways. This property of distributed signal representations is most useful for machine intelligence, imagination and creativity.

Distributed signal representations have a drawback. Each feature signal gets its basic meaning from the elementary feature detector. Therefore the basic meanings of the feature signals are tied to the hardware. Consequently, distributed signal patterns cannot be transported into other systems, unless these systems have exactly similar hardware. This issue is related to the detection problem of qualia, see chapter 4. Binary numbers and code words are different, because they are based on a general definition. Therefore they can be transported to other systems that comply with the general definition.

Summary

- Pavlovian conditioning and Hebbian learning are simple associative mechanisms.
- Associative information processor operates with distributed neural signal representations, is able to learn correlations, utilizes associative linking and evocation, facilitates associative memories and enables natural transition from sub-symbolic to symbolic processing.
- Distributed signal representations depict entities by their features and are suitable for the detection of content and meaning.
- A distributed signal representation is an array (vector) of one ("a grandmother signal") or more signals.

Chapter 12

Neural Realization of Associative Processing

12.1. Spiking Neurons or Block Signal Neurons?

The biological neuron is a spiking neuron. When excited, it produces trains of equal intensity electric pulses or spikes with a varying repetition rate. Obviously, both the presence of the signal and its pulse repetition rate are carrying information. It is also possible that in some cases also the timing of the individual spikes may convey some information. A simple biological neuron model is depicted in Fig. 12.1.

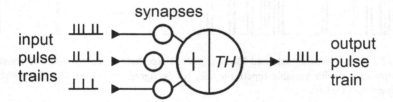

Fig. 12.1. The simple biological neuron model receives input pulse trains via synapses, sums their energies and produces an output pulse train if the sum exceeds the excitation threshold *TH*.

In simple neuron models, the neurons receive incoming pulse trains via so-called synapses, multiply the input signal pulse strength by the synaptic strength, sum the energy of the individual pulses over a length of time and fire an output pulse train of their own, if the sum of the received spike energies exceeds a set excitation threshold *TH*. This simple neuron model is also known as the *integrate and fire* model, introduced originally by Louis Lapique already around 1907 [Abbott, 1999]. Since then a number of more complicated models for spiking neurons have been devised.

Complex neuron models are necessary, when the various biological neuron-related phenomena are to be explained. However, simple models may suffice, when only the logic of the information process is considered. This is not unlike the case of the electron tube; for the complete explanation of the operation of the electron tube one must explain also the operation of the filament (quite complicated really!), but in the explanation of actual electron tube applications, the explanation of the complex processes of the filament is irrelevant. After all, electron tubes can be replaced by transistors, which do not have a filament at all.

In artificial neural networks it is not necessary to simulate and copy the metabolism of biological synapses and neurons. Instead, the emulation of the functional aspects that relate to the actual information processing is sufficient. This applies to the signal forms, too. Therefore, instead of mimicking the biological pulse train signals, more simple block signals may be used, Fig. 12.2.

Fig. 12.2. Pulse train signals and block signals. Pulse train signals consist of constant intensity spikes with variable repetition rate. Block signals may have variable duration and variable intensity.

The presence of the block signal can convey the same information as the presence of the corresponding pulse train. The pulse repetition rate in a pulse train may convey additional information such as the urgency or importance of the signal. In block signals this information may be encoded in the intensity or voltage value of the signal. The leading edge of a block signal may be used to convey timing-related information. Thus, the information carrying capacities of pulse trains and block signals would seem to be similar. In practical terms, the block signals are easier to implement and therefore are used by the Haikonen associative neuron.

12.2. Associative Neurons and Synapses

The associative neural processing of distributed signal representations calls for special neurons. The Haikonen associative neuron is one such neuron, which is also able to realize the directness requirements of qualia and match/mismatch/novelty detection [Haikonen 1999, 1999b, 2007].

The operation of the Haikonen associative neuron is generalized from the requirements for Pavlovian conditioning and Hebbian learning. In Pavlovian conditioning two signals are associated with each other and consequently, a neuron that is able to execute basic Pavlovian conditioning could be very simple. In more general and demanding cases one signal must be associated with a number of simultaneous signals, also known as a signal array or vector.

The Haikonen neuron is designed to execute the association of one signal with a signal vector. The single signal to be associated is called here the *main signal* and the signal vector is called the *associative input vector*. Thus the neuron has one main signal input, one main signal output and a number of associative synaptic inputs that correspond to the individual signals of the associative input vector, see Fig. 12.3.

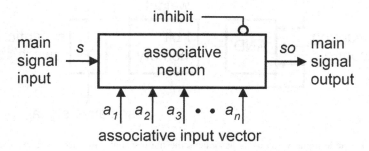

Fig. 12.3. The Haikonen associative neuron associates one main signal with an array of associative input signals (an associative input vector). This association takes place via learning and after learning the associative input is able to evoke the main signal as the output also in the absence of the main signal at the input.

The Haikonen neuron conserves the meaning of the main signal; the meaning of the main input and output signal is the same. This property satisfies the minimum aspect of the directness requirement of qualia.

The Haikonen associative neuron is a learning memory unit. Initially there are no connections between the associative inputs and the main signal output and consequently any signals at the associative inputs will not evoke any signal at the main signal output. The eventual connections between the associative inputs and the main signal output will be learned from the coincident occurrences of the main signal and the associative signals.

Within the neuron, the circuit unit that registers the coincidences of the main signal and an associative signal, is called *the synapse*. There is one synapse for each associative input signal. The synapse consists of a coincidence detector, a one bit memory and a synaptic switch, Fig. 12.4.

The content value of the one bit memory is called the *synaptic weight* of the synapse and it controls the so called *synaptic switch*. The synaptic weight can have only two values, zero and one. When the synaptic weight value is zero, the synaptic switch is open and the associative signal does not pass the synapse. When the synaptic weight value is one, the synaptic switch is permanently closed and the associative signal passes the synapse.

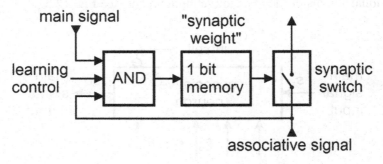

Fig. 12.4. The associative synapse associates the main signal with one associative signal. Association takes place via learning whenever the main signal and the associative signal occur at the same time. This sets the synaptic weight to the logical value 1. This in turn closes the synaptic switch and the associative signal gets a permanent path into the neuron.

The operation of synapses can also be considered as multiplication; each associative signal of the associative input vector is multiplied by its corresponding synaptic weight, zero or one. The sum of the products of the associative signals and their synaptic weights is called here *the synaptic excitation sum*. The associative input vector evokes the output signal of the neuron, whenever the synaptic excitation sum exceeds a set threshold value.

The associative synapse of the Fig. 12.4 executes simple Hebbian learning that is based on the simultaneous occurrence of the main signal and the associative signal. The synaptic weight value is determined by the logical AND circuit, which acts as the coincidence detector that registers the coincident occurrences of the main signal and the associative signal.

Initially, the content of the one bit memory is zero and the synaptic switch is open. Next, whenever the main signal and the associative signal have the value of logical one at the same time, the output of the AND circuit will turn to logical one and the content value of the one bit memory, the synaptic weight, will be set to one. This will close the synaptic switch and the associative connection is thus created. In this case already one coincident occurrence is sufficient for the creation of the associative connection between the main signal and the associative signal.

In many cases some kind of learning control is necessary as the learning in given groups of neurons must take place only at suitable occasions determined by the focus of attention. This is facilitated by the learning control signal, which has to have the logical value of 1 in order to allow learning.

The complete associative neuron is depicted in Fig. 12.5. The complete associative neuron circuit consists of a number of synapses, a summing circuit for the associative input signals that pass the synaptic switches and an output threshold circuit. The threshold value can be externally controlled.

In some applications the main signal must pass the neuron. In those cases the input main signal may be summed to the synapse outputs and forwarded to the threshold circuit. In that case an additional sum circuit is used as shown. The main signal will pass the neuron depending on the set threshold value. In all applications the meaning of the main signal is carried over to the output.

An associative vector that is associated with the main signal, or a somewhat similar vector, may evoke the output signal. The associative signals that pass the synapses are summed together and the resulting signal value is passed to the output threshold circuit. If the sum value of the output signals (the synaptic excitation sum) from the synapses is higher than the set threshold value, then the output main signal will have a positive value, otherwise the output main signal will have the value zero. The threshold value is variable and is determined by the interconnection with other neurons.

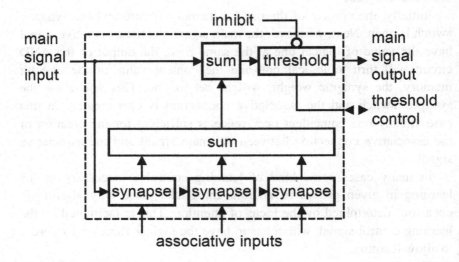

Fig. 12.5. The complete associative neuron circuit consists of a number of synapses, a summing circuit for the associative input signals that pass the synapses and an output threshold circuit. The threshold value can be externally controlled. In some applications the main signal is passed through the neuron. In that case an additional sum circuit is used as shown.

The so-called Winner-Takes-All (WTA) output threshold circuit is used, when several neurons are used as a group and only the neuron (or those neurons) that has the highest synaptic excitation sum is allowed to send its output signal. Figure 12.6 presents the principle of the Winner-Takes-All output threshold circuit.

In Fig. 12.6 inputs 1, 2 and 3 are the synaptic excitation sum values from the corresponding neurons 1, 2 and 3. These values are compared by the comparator to the threshold value (TH), which is common to the all neurons of the group. If the input value is higher than the TH value, then the TH value is replaced with the input value. At the same time the output switch is closed and consequently the output will now equal the input. At other times the output will be zero.

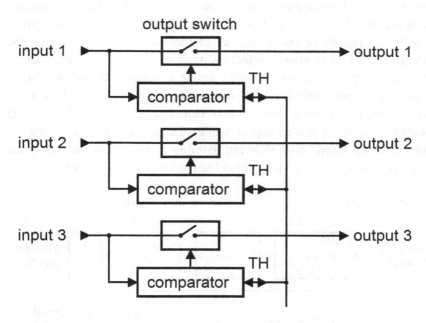

Fig. 12.6. Winner-Takes-All (WTA) output threshold is used when an output from a number of neurons is to be selected. The comparator compares the input to the TH value and if the input is higher, then the TH value is replaced with the input value and at the same time the output switch is closed; the output will now equal the input. At other times the output will be zero.

In Fig. 12.6 the TH connection is bi-directional, that is, each threshold circuit may receive or transmit a value. Normally the TH value is received, but if the synaptic excitation sum of a given neuron is higher than the instantaneous TH value, then the neuron transmits this higher value to the TH-line. Only one controlling TH-line is required regardless of the number of interconnected neurons. An actual realization example of a WTA circuit is given by Haikonen [2007].

12.3. Correlative Learning

Simple Hebbian learning is fast, but it has its shortcomings. Consider for instance a situation, where the adjective word "blue" is to be defined. One may show blue objects to the associative system and try to associate the word "blue" with the feature of "blue". However, the feature "blue" is only one feature of the whole set of the features that constitute the appearance of the given object. Consequently, any word that is associated with the feature set, will become a name for the whole object. Therefore a correlative learning method is required, which is able to utilize several example objects and can associate a word with the common feature of the examples while rejecting associations with the other features. Correlative Hebbian learning fulfills these requirements. A synapse using correlative Hebbian learning is depicted in Fig. 12.7.

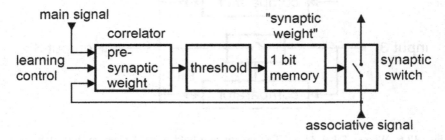

Fig. 12.7. A synapse using correlative Hebbian learning. The pre-synaptic weight goes up a little at each coincidence of the main signal and the associative signal and goes down a little if the main signal appears alone. If a correlation between the main signal and the associative signal exists, the pre-synaptic weight will exceed the threshold and the actual synaptic weight will be set to "one". Thereafter the correlator circuit has no longer any function.

In Fig. 12.7 the AND-circuit of the simple synapse of Fig. 12.4 is replaced by a correlator circuit that accumulates the so-called pre-synaptic weight value. The pre-synaptic weight goes a small step up at each coincidence of the main signal and the associative signal. If the main signal appears alone, the pre-synaptic weight goes a small step down. If a correlation between the main signal and the associative signal exists, the pre-synaptic weight will eventually exceed the preset threshold and the actual synaptic weight will be set to "one". Thereafter the correlation circuit has no longer any function.

In the following the learning of the association of the concept "blue" (for instance, the word "blue") with the feature <blue> is presented as an example. In this simplified case the single main signal of the neuron represents the concept "blue" while a number of associative feature signals represent the various features of perceived objects. Let us assume, that three objects are used for the training; these objects are a blue book, a blue flower and a blue pen. All these objects activate the common feature "blue" and also a number of other features that are specific to each individual object. In this example the other features are represented by one signal for clarity, see Fig. 12.8.

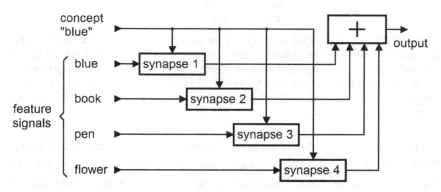

Fig. 12.8. The association of the concept "blue" with the sensory feature <blue> by using the examples of blue pen, blue book, and blue flower.

Figure 12.8 depicts a neuron with four synapses. Neurons are initially wired in a general way that allows the learning of various associative connections. In this example the main signal is set to represent the concept "blue" while the meaning of the feature signals are determined by the fixed wiring from the feature detectors. The task is to associate the feature <blue> with the concept "blue" and reject all other associations.

Thus, only the feature <blue> should evoke the concept "blue" at the output of the neuron, while the features of <book>, <pen> or <flower> should not evoke any output. This will happen, if the synaptic weight of the synapse 1 has the value 1 and the other synaptic weights have the value 0. Correlative Hebbian learning picks up the common feature among a number of objects and works here as follows, see Fig. 12.9.

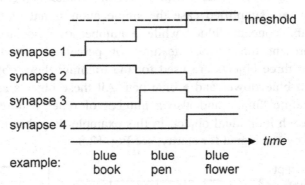

Fig. 12.9. Pre-synaptic weight values during the correlative learning of the association of the concept "blue" with the feature <blue> using the examples of a blue book, a blue pen and a blue flower. After the third example object, "blue flower", the pre-synaptic weight of the synapse 1, blue/blue, exceeds the threshold and the concept "blue" and the feature <blue> are associated with each other. Other pre-synaptic weights will eventually fade away.

During training the blue book is presented first and the concept "blue" is given. At this moment the pre-synaptic weights for the blue/blue synapse 1 and blue/book synapse 2 step up and others remain at zero. Next, the blue pen is presented. Now the blue/blue synapse 1 steps up again and also blue/pen synapse 3 steps up while others go down. After the last example object, the blue flower, the blue/blue pre-synaptic weight has the highest value and the learning may be completed by using a threshold value of 2.5. The synaptic weight of the synapse 1 will now have the value 1 and the other synaptic weight values will remain at the value 0; the neuron has learned the association between the feature <blue> and the concept "blue" and thereafter the presence of the feature <blue> will evoke the corresponding concept "blue" signal at the output of the neuron.

Likewise, the common features of various examples of books could be associated with the concept of "book", while the associations with untypical features, which might be common to totally unrelated objects, would be rejected. In this way Correlative Hebbian learning is useful in the learning of simple meanings of words.

12.4. The Associative Neuron as a Logic Element

Is it possible to compare the associative neuron to a logic element such as the logic AND or OR, and if so, what would be the executed logic function? If the associative neuron executed a standard logic function, then would it be possible to build associative neural networks with the readily available digital microchips? A simple analysis gives the answer.

The output condition for an associative neuron with three associative input signals can be written as follows:

$$so = 1 \text{ IF } a_1*w_1 + a_2*w_2 + a_3*w_3 > threshold, \text{ ELSE } so = 0$$

where

so = output signal, (1 or 0)
a_i = associative input signal, (1 or 0)
w_i = synaptic weight, (1 or 0)

Let us assume that all the synaptic weights w_1, w_2, w_3 have the value 1. This would mean that the associative input vector $[a_1, a_2, a_3] = [1, 1, 1]$ has been associated with the main signal. The condition for the output becomes now:

$$so = 1 \text{ IF } a_1 + a_2 + a_3 > threshold, \text{ ELSE } so = 0$$

If the *threshold* is set to 2.5, then *so* will become 1 only if every a_i has the value 1. This corresponds to the logic AND function:

$$so = a_1 \text{ AND } a_2 \text{ AND } a_3$$

If the *threshold* is set to 1.5 then *so* will become 1 only if at least two of the a_i have the value 1. This corresponds to the logic function:

$$so = (a_1 \text{ AND } a_2) \text{ OR } (a_2 \text{ AND } a_3) \text{ OR } (a_1 \text{ AND } a_3)$$

The accepted associative input vectors would be: [1,1,0], [0,1,1], [1,0,1], [1,1,1]. This can be considered as a kind of a "SOFT-AND"; all inputs of the true logic AND are not necessary. This means that an associative input vector that is an approximation of the original associative input vector, is also able to evoke the associated main signal. This is a useful property in many practical applications as it categorizes associative inputs and automatically applies learned associations to similar novel situations. SOFT-AND allows also some random variation in the associative input vector.

If the *threshold* is set to 0.5 then *so* will become 1 only if at least one a_i has the value 1. This corresponds to the logic OR function:

$$so = a_1 \text{ OR } a_2 \text{ OR } a_3$$

OR function is also useful in many practical applications.

Thus, it can be seen that by varying the *threshold* value the associative neuron can be made to execute different logic functions according to the requirements of each application. For the sake of clarity and simplicity an associative vector with three signals was used in the previous examples, but it should be obvious that these principles apply to associative vectors of any length.

Thus, it can be determined that in those applications, where the true AND and true OR are to be executed, standard logic circuits could be used. There are no standard microchips for the SOFT-AND function. However, the associative neuron has also the learning function that is implemented by simple Hebbian synapses or the more complex correlative Hebbian synapses. Standard microchips that execute the Hebbian learning function and the logic function are not currently available.

12.5. Associative Neuron Groups

12.5.1. *The Operation of an Associative Neuron Group*

Associative neuron groups are used for the evocation of one signal from many possible ones. An associative neuron group consists of a number of neurons. The associative inputs of these neurons are connected together so that at each moment the same associative input vector appears at the associative inputs of each neuron. Each associative input vector may be associated with one main signal. This main signal may be evoked if an associative input vector matches closely the originally associated vector. The output thresholds of the neurons are connected together in a way that allows the "Winner Takes All" operation. This operation will allow the outputting of the main signal that has been evoked by the closest matching associative vector. The structure and depictions of an associative neuron group is presented in Fig. 12.10.

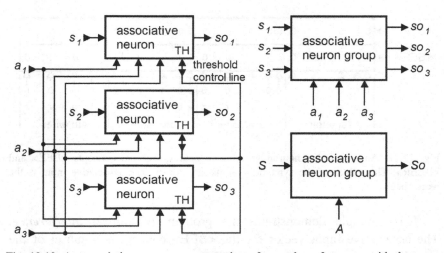

Fig. 12.10. An associative neuron group consists of a number of neurons with the same signal vector as the associative input for each neuron. Left, detailed depiction. Right, two alternative depictions.

An example illustrates the operation of the neuron group. Let's assume that we have two concepts that are <BEE> and <BEER>. These concepts are depicted by one signal each. The task is to evoke associatively the corresponding concept signal when the written name of the concept is given. For this purpose two neurons are used, one for the <BEE> concept and another for the <BEER> concept. The associative input signals consist of the letter signals. The first, second, third etc. letter of a word is one of the alphabets, therefore the number of the associative input lines will be the maximum allowable number of letters in a word times the number of alphabets. This configuration is depicted in the Fig. 12.11.

The evocation strength for each neuron is computed as the sum of active associative signals multiplied by the corresponding synaptic weights. In this example the evocation strength for the <BEE> concept will be 1x1 (first letter) + 1x1 (second letter) + 1x1 (third letter) = 3. Likewise, the evocation strength for the <BEER> concept will be, surprisingly, the same = 3. Which one of these concepts should be evoked?

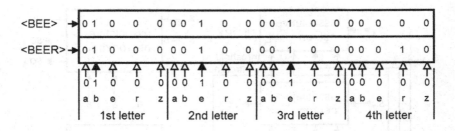

Fig. 12.11. An associative neuron group with neurons for the concepts <BEE> and <BEER>. The values of the synaptic weights are shown. The associative input is the word "bee".

This example demonstrates the problem of *sub-set interference*. The associative input vector for the <BEE> concept is a subset of the associative input vector for the <BEER> concept and consequently the straight forward computation of the evocation strength will not yield any difference between the evocation strengths. However, the problem can be remedied by a simple modification of the evocation strength computation rule. It can be defined that the multiplication of zero associative input by the synaptic weight value of one gives the value of -1 ($0 \times 1 \Rightarrow -1$). Thus, in this case the evocation strength for the <BEER> concept would be 1×1 (for b) + 1×1 (for e) + 1×1 (for e) + -1(for r) = 2. This is less than the evocation strength 3 of the <BEE> concept and consequently the correct decision can be made by applying an output threshold value of 2.5, which is exceeded only by the <BEE> signal.

More generally, the value of the output threshold should be floating and slightly less than the maximum evocation strength value. This would implement the *Winner-Takes-All* threshold operation. (Detailed discussion about the operation of associative neuron groups and their interference modes is given in [Haikonen 2007]).

12.5.2. *The Association of Vectors with Vectors*

The associative neuron group is able to associate an associative input vector with one main signal without interference. Also one associative signal can be associated with a main signal vector without interference. When associative signal vectors are to be associated with main signal vectors without error, the arrangement of the Fig. 12.10 can be used. Also other arrangements are possible.

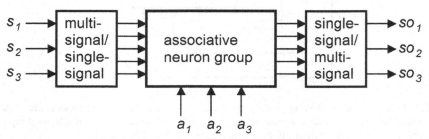

Fig. 12.12. The association of vectors with vectors.

In Fig. 12.12 the main input vector $\{s_1, s_2, s_3\}$ is transformed into a single signal representation and the associative input vector $\{a_1, a_2, a_3\}$ is associated with one of the single signals. At the output of the associative neuron group the single signal representation is transformed back into the vector form. Thereafter an associative input vector $\{a_1, a_2, a_3\}$ will evoke the associated output vector $\{so_1, so_2, so_3\}$, which is similar to the input vector $\{s_1, s_2, s_3\}$ that was used in the training.

12.5.3. *Autoassociative Memory*

The associative neuron group can be wired to work as an autoassociative memory. In this application the main signal vector is used also as the associative vector, Fig. 12.13. The main signal vector S will thus be associated with itself. The autoassociative memory utilizes the previously described SOFT-AND function. This function allows the evocation of the whole input vector S as the output vector So when a small arbitrary subset, even with some distortions, is introduced as the input. The principle of autoassociation in various forms is useful in the realizations of associative cognitive architectures. The autoassociative memory is useful in applications where the main input signal vectors are very large.

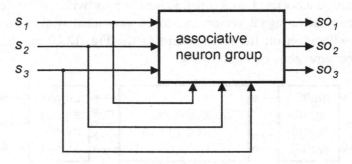

Fig. 12.13. An associative neuron group wired as an autoassociative memory. Here the associative neuron group associates the input pattern with itself. A part of a learned input pattern will evoke the full original pattern via the SOFT-AND function.

12.5.4. *Temporal Sequences*

Associative neuron groups can be used to associate simultaneously present signals and signal vectors with each other. In actual cognition also temporal succession occurs; events follow each other and must be learned and remembered as such. Not only the temporal order is important, many times also the duration of the events count and must be remembered. For instance, a spoken word is a sequence of phonemes of different durations. Moreover, these durations are relative, a word can be spoken slowly or fast, yet it must be recognized as the same word. Thus, the processing of temporally sequential events calls for means for the association of non-simultaneous signals and signal vectors with each other and also means for the recording and replay of the absolute and relative durations of the individual events.

The basic associative neuron and neuron groups cannot associate non-simultaneous signals and vectors with each other. Therefore, the temporally successive, serial signal patterns must be transformed into temporally parallel forms, where all the individual signal patterns are available at the same time. This can be done by short-term memories and delay lines. Figure 12.14 depicts the principle of a circuit that associates a short sequence with a cue vector.

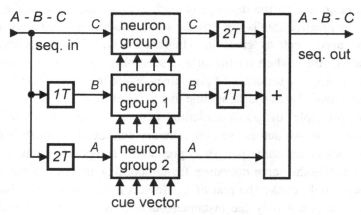

Fig. 12.14. An example of an arrangement for the association and associative evocation of a sequence with a cue vector. During learning the cue vector is associated with three successive *A, B, C* vectors of the input sequence. During replay the cue vector evokes these successive vectors at the same time. However, they are output serially due to the *1T* and *2T* delay lines.

In Fig. 12.14 the input sequence *A-B-C* consists of individual *A, B* and *C* signal vectors, groups of distributed feature signals that appear serially. The cue vector is a signal pattern that is to be associated with the whole input sequence, so that later on it can be used to evoke the *A-B-C* sequence as the output. Three neuron groups are used in this circuit. The neuron group 0 associates the C-vector of the input sequence with the cue, the neuron group 1 associates the B-vector of the input sequence with the cue and the neuron group 2 associates the A-vector of the input sequence with the cue. The units $1T$ and $2T$ are delay lines that delay the input sequence so that the A, B and C vectors appear simultaneously at the inputs of the neuron groups. The actual association takes place, when the last vector, the vector C, of the sequence is received.

During the replay of the sequence the cue vector evokes these successive vectors at the same time. However, they are output serially due to the $1T$ and $2T$ delay lines. the vector A is output first, then the vector B appears after the $1T$ delay and finally the vector C appear after the $2T$ delay. The summing unit passes only the most recent vector so that the A, B and C vectors will not overlap.

The circuit of Fig. 12.5 operates properly, when the sequence has fixed timing, that is, the duration of each event is the same as the delay time T. In practice this is not usually so and additional means of timing must be used, such as given in [Haikonen 2007]. The evocation of a sequence is used, when for instance, a seen object evokes its name. The name is a temporal sequence of phonemes, while the visual features of an object are available at the same time in a parallel way.

The principle of autoassociation can be extended to temporal sequences, too. An autoassociative memory can recall complete patterns when a part of the same pattern is presented. Temporal autoassociation would execute the same operation for temporal sequences; a part of the sequence would evoke the rest of it. However, the problem is that in a temporal sequence only the instantaneously present part is available at each moment and therefore no association can be made with any past or future part of the sequence. The solution to this problem is, of course, the use of memory. With the help of temporally ordered memories, the present part of a sequence can be associated with the temporally previous ones. After the learning of these associations, the temporally previous patterns will be able to evoke those patterns that follow.

Chapter 13

Designing a Cognitive Perception System

13.1. Requirements for Cognitive Perception

Without perception there cannot be any consciousness. Without perception processes an agent cannot acquire information about and become aware of the external world, the agent itself and its own mental content. Therefore perception is a necessary function for any cognitive architecture that is intended to be conscious to any extent.

It was argued before that conscious perception calls for qualia and the basic requirements for qualia include the direct and transparent representation of information that also preserves the amodal features of the perceived phenomena. These requirements cannot be satisfied in symbolic systems, like those that digitize sensory information and represent it by numbers. However, these requirements can be accommodated in sub-symbolic systems, like those that utilize distributed representations.

Consequently, the design of a cognitive architecture should begin with an assessment of the requirements of sensory and introspective perception. The perception process for artificial cognition is presented in Fig. 13.1.

A cognitive agent may have a variety of sensory systems, sensory modalities, that detect various physical phenomena, such as light, sound, touch, chemicals (smell and taste), temperature, etc. As discussed before, these phenomena as such are not suitable forms for information processing in the cognitive machinery. One cannot input photons, vibrations of air pressure or chemicals directly into a computer or an artificial neural network. Therefore the first step for perception is transduction. This transduction is performed by a sensor, which transforms the sensed phenomenon into the common internal form that is used inside the cognitive machinery. In artificial electronic systems this common form would be that of electric signals.

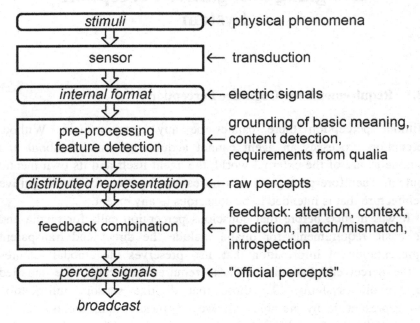

Fig. 13.1. Perception process for artificial cognition.

The next requirements arise from the style of the representation of information. Associative information processing with distributed signal representations calls for pre-processes that extract feature signals from the sensory information. These pre-processes would be based on feature detectors. These detectors would also provide the grounding of basic meaning for the feature signals that constitute the distributed signal arrays; the presence of a feature signal would indicate that the corresponding feature has been detected by the specific feature detector. Feature detection would also satisfy the basic requirements for content detection.

The distributed representations from the feature detectors constitute kinds of raw percepts. However, perception is not a simple stimulus-response process, it is also affected by the internal state of the cognitive system. Therefore the raw percepts must be subjected to the effects of the internal factors such as attention, context, prediction and the like. These factors would relate to the selection of context related information, augmentation of information by experience, searching for certain information, effects of expectation and so on. This action calls for a point of re-entry for the relevant internal signals. This re-entry is called here feedback. At the re-entry point the sensory feature signal patterns can be compared with the feedback signal patterns; this will satisfy the requirement of the detection of match/mismatch/novelty conditions.

The distributed signal arrays that are modified by the feedback satisfy the basic requirements for perception and are thus the "official" output of the perception process. Here these signal patterns are simply called *percepts*. These percepts are forwarded or broadcast to the rest of the cognitive machinery.

The requirements for the perception of physical phenomena and for introspection, the perception of internal mental states, might seem to be completely different, but in practice the situation may not be so (see Sec. 7.3). Introspection always shows mental entities that are in the form of the entities of some sensory modality. Inner speech is in the form of sound patterns, inner imagery is in the form of visual patterns, other mental entities may be in the form of motion patterns and feel patterns, etc. Thus, introspected mental entities are such that could be sensorily perceived. The feedback transforms inner signal patterns into sensory signal patterns and therefore the introspected mental entities would appear as sensory percepts, but not necessarily with the full and vivid details of actually perceived objects. Thus the feedback also allows the utilization of the perception process for the introspection of inner mental content, too.

The perception/response feedback loop is one circuit arrangement that realizes the above requirements for cognitive perception and also provides additional functions such as short-term memory and the transition from sub-symbolic to symbolic processing by the association of additional meanings with the percept signal patterns.

13.2. The Perception/Response Feedback Loop

13.2.1. *The Basic Principle*

The general requirements of perception can be satisfied by a feedback (also known as re-entry) circuit arrangement, which is here called the perception/response feedback loop. This circuitry receives information from a sensor and also from the output of the internal neural processes in the form of feedback signals. The internal percept arises from the combined effect of the sensory and feedback signals. The principle of the perception/response feedback loop is depicted in Fig. 13.2.

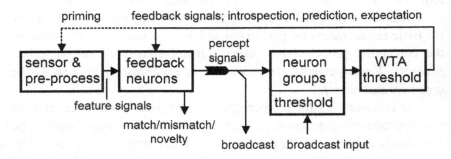

Fig. 13.2. A simple perception/response feedback loop. Pre-processed feature signals are forwarded to the feedback neuron group, which also receives signals from the output of the internal neuron groups. The output signals from the feedback neurons are called the percept signals and are broadcast to other units. Here each line depicts a number of parallel signal lines.

Figure 13.2 depicts the general structure and principle of the perception/response feedback loop. Information acquisition is performed by a sensor. The pre-processing unit pre-processes the sensory information into an array (vector) of elementary feature signals. Each elementary feature signal depicts the presence of the corresponding sensory feature and the intensity of the signal depicts its importance.

The feature signals from the sensory pre-process are forwarded to the feedback neurons. The feedback neurons receive information also from the internal processes of the system. This information is in the form of feedback signals and may act as an expectation, prediction or an introspection. Match/mismatch and novelty signals are generated on the basis of the relationship between the feedback and the sensory information. The output signals from the feedback neurons are called the percept signals.

A complete cognitive system utilizes a number of perception/response feedback loops, which are associatively cross-connected to each other. The percept signals are broadcast to the other perception/response feedback loops and are also forwarded to the internal neuron groups. These neuron groups receive broadcast inputs from the other perception/response feedback loops. The input thresholds of the neuron groups determine, which broadcasts, if any, are accepted at each moment. The neuron groups are able to learn and remember associative connections between the percept signals and the received broadcast signals. The neuron groups are also able to learn and remember temporal sequences.

The neuron groups may generate a large number of signals as candidates for the output, but only one group of signals can be accepted at each moment. The Winner-Takes-All threshold (WTA) at the output executes the selection and passes only the strongest candidate signals at each moment. These output signals will be returned to the feedback neurons as the feedback signals.

The feedback loop is able to re-circulate and sustain percept signals for a while after the actual sensory stimulus has disappeared. Therefore the percept points are also *perceptual short-term working memory locations*.

13.2.2. *The Realization of Priming*

The so-called *priming effect* occurs, when previously learned percept patterns affect the perception process. The priming effect can occur in different sensory modalities. The priming effect can be positive or negative. Positive priming helps to recognize patterns during noisy conditions, while negative priming may prevent recognition. For example, a person may be searching for a certain object and may fail to see it, because the priming has provided expectations of a different-looking object. Primed perception is depicted in Fig. 13.3.

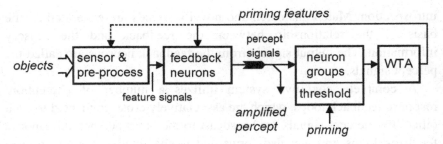

Fig. 13.3. Primed perception. Percept signals that match the priming signals are amplified.

In the perception/response feedback loop the priming effect is facilitated by the feedback signals. During positive priming the priming features more or less match the sensory feature signals and the combined signals will be stronger than the non-primed signals. Therefore the primed signals will win other signals and will be forwarded as the percept signals. During negative priming the priming does not match the actually desired sensory features and no signal amplification takes place for these features. The priming signals may be evoked internally by signal patterns from other perception/response feedback loops. Also a part of the sensory percept may evoke autoassociative expectations about the whole sensory pattern. Priming may be extended to the actual sensor pre-process.

13.2.3. *Match, Mismatch and Novelty Detection*

In the perception/response feedback loop there are three possible relationships between the sensory feature signal vector and the feedback signal vector. These relationships can be resolved at the feedback neuron group and they are:

- The *match condition*: the sensory feature signal vector is similar or almost similar to the feedback signal vector.
- The *mismatch condition*: the sensory feature signal vector is not similar or almost similar to the feedback signal vector.
- The *novelty condition*: No feedback signal vector is present.

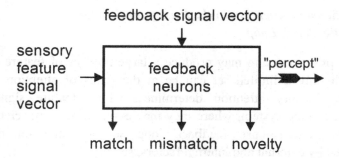

Fig. 13.4. The feedback neuron group generates match, mismatch and novelty signals according to the relationships between the sensory feature signal vector and the feedback signal vector.

Figure 13.4 depicts the match, mismatch and novelty signal outputs from the feedback neuron group. Binary signals can be used; during the match condition the match signal is set to one, other signals are zero. During mismatch the mismatch signal is set to one and other signals remain at zero. During novelty condition no feedback signal vector is present while a sensory feature signal vector is present. In this case the novelty signal is set to one and the other signals remain at zero. When no sensory signals and feedback signals are present, the match, mismatch and novelty signals remain at zero.

Complete match between the sensory feature signal vector and the feedback vector may be rare, therefore the decision between the match and mismatch state must be based on the relative number of individual signal matches and mismatches.

It should be noted that the signals in the sensory feature signal vector and the feedback signal vector represent elementary features and therefore can be compared with each other. If, for instance, these vectors were to represent pixel maps, then the comparison would not be feasible; match condition would rarely occur even in those cases, where the images represented by the pixel maps were quite similar.

Match, mismatch and novelty detection is necessary for various cognitive tasks, such as prediction, attention control, search operation, answering *yes* or *no* to questions like "is this xx?". Match, mismatch and novelty conditions are also related to emotions.

13.2.4. *Sensory Attention in the Perception/Response Feedback Loop*

Sensory pre-processing may produce a large number of feature signals, of which only a limited set can be at the focus of attention at each moment. Sensory attention determines, which feature signals are forwarded to the system, where they may evoke further associations. In the perception/response feedback loop sensory attention may be controlled by external and internal factors, see Fig. 13.5.

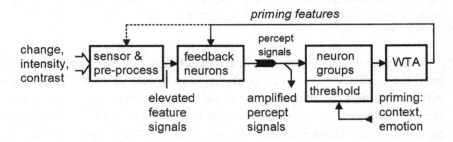

Fig. 13.5. Attention in the perception/response feedback loop may be guided externally by change, high intensity and contrast in the sensed entity. Internally attention may be controlled by priming. Attended percept signals are stronger and may therefore win other percepts elsewhere in the system.

External sensory attention guiding factors include novelty and sudden changes in the sensed modality, higher intensity and contrast. For instance, in the auditory modality this would mean sudden unexpected sounds, changes in ongoing sounds, loud sounds and sounds that are very different from other sounds. In visual modality attention would be captured by bright objects, objects with high visual contrast and by changing and moving objects. The sensory feature pre-processor should be designed in such a way that these properties would cause elevated signal levels for the related feature signals. This in turn would help these signals to pass the self-adjusting output threshold of the feedback neuron group. Thereafter these signals would then appear as amplified percept signals.

Internal sensory attention guiding factors include context and emotional significance. These factors operate associatively via the neuron groups inside the perception/response feedback loop. Context and emotional significance may evoke certain feature signals that are fed back to the feedback neurons as priming features. This leads to the amplification of the primed feature signals and the selection of these by the output threshold circuit. The primed feature signals would then appear as amplified percept signals.

Percept signals are broadcast to the rest of the system. There they will have to compete against other broadcasts at various threshold circuits. The amplified percept signals will win the other signals more likely and therefore may become the focus of global processing and in this way may win the global attention of the system.

13.2.5. *The Realization of Visual Searching*

During a visual search a given object is looked for. The visual modality perception/response feedback loop facilitates the search operation easily, Fig. 13.6.

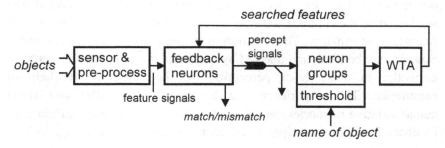

Fig. 13.6. Search operation. The name of the searched object evokes the visual features of the object. These are fed back to the feedback neuron group. When the searched object is found, the feedback matches the perceived object and a match-signal is generated.

An "inner image" of the searched object is evoked in the visual modality perception/response feedback loop by, for instance, the name of the object. This "inner image" is not to be seen as a pixel map image of the object, instead it consists of a set of more or less invariant features that might match the searched object generally. These searched features are forwarded to the feedback neurons, where they may or may not match the sensory features of the seen objects at the given instant. When the searched object is found, the sensory features will more or less match the feedback features and a match-signal will be generated. This signal indicates that the search has been successful and can be stopped. Otherwise a mismatch-signal will be generated and the search will continue.

It should be noted that in associative processing also more general visual search methods exist. For instance, one may be in a foreign city looking for a nice restaurant for lunch. In this case one does not have any pre-existing inner image of the searched restaurant. Instead, new visual percepts may or may not evoke ideas of a restaurant and some locations may remind of a restaurant. This mode of operation calls for the cooperation of several modules.

13.2.6. *Predictive Perception*

Sensory perception is not always a simple process of detecting what is out there. In the visual world objects may be partially covered and shadowed or the illumination may be bad, all this leading to the loss of important information. The situation in the auditory perception is similar. Sounds may be weak and masked by noise and therefore hard to perceive correctly. In these cases perception can be aided by the help of experience. This experience can be stored in parallel and serial autoassociative memories, which can be added to the perception/response feedback loop. The principle of this arrangement is shown in Fig. 13.7.

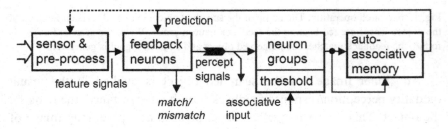

Fig. 13.7. The addition of an autoassociative memory allows prediction in the perception/response feedback loop. *Match/mismatch* signals indicate the success of prediction.

In Fig. 13.7 an autoassociative memory is added to the perception/response feedback loop. The input of this memory consists of the percept signals or the signals that are evoked at the neuron groups in front of the associative memory. The output of the associative memory consists of the evoked signal patterns, or in the default case, when nothing is evoked, the input signal patterns of the memory. This output is returned via the feedback loop back to the feedback neurons. At the feedback neurons the predicted signals can be compared to the actual sensed feature signals. The result of this comparison can be indicated by match/mismatch signals.

In the visual modality a parallel autoassociative memory would enable the prediction of the unseen parts of a known object. A temporally serial autoassociative memory would allow the prediction of motion, the next position of a moving object, when the object moves in ways that are already known to the system.

In the auditory modality a serial autoassociative memory would be used. This would allow the prediction of the continuation of the sound. In linguistic systems the next word could be predicted, and as this prediction is returned via the feedback loop to the percept point, this would again evoke the next word and so on; the whole string of words would be recalled. (The autoassociative prediction of word strings like one-two-three-four-five... in a perception/response feedback loop is demonstrated in [Haikonen 1999].)

With the inclusion of autoassociative memories the perception/response feedback loop can support simple sensory prediction, as described above. *Cognitive prediction* is more complicated and involves the coalition of the experience of several modules. Examples of cognitive prediction would be, for instance, the prediction of weather: "Black clouds are gathering in the sky; what may happen next?" or the prediction of future: "What will happen, if I spend more money than I earn?".

13.2.7. *Conscious Introspection of Imaginations*

Imagination is the mental perception and manipulation of actions and entities that are not physically present. Imagination utilizes similar representations that result from direct sensory perception, but in this case these representations arise from and are modified by internal causes. However, internal neural activity is not observable as such and is in this way "sub-conscious", because, as stated afore, only percepts can be consciously observed. Therefore, the realization of imagination in a cognitive system calls for mechanisms that enable introspection via the perception of some of the inner neural activity as virtual percepts of various sensory modalities.

The perception/response feedback loop module provides a mechanism for the translation of inner representations into percepts and in this way it facilitates conscious introspection. The feedback loop works as a pathway that leads internal neural activity patterns back to the perception process. When sensory signals are attenuated, the feedback signals will dominate and will become the official percepts. Percept signals have qualia, therefore also the internal content will be perceived in terms of sensory qualia. These imaginary percepts would be emotionally evaluated and broadcast to other modules just like normal percepts. The other modules, in turn, would generate responses and broadcast those. In this way loops of activities would follow, Fig. 13.8.

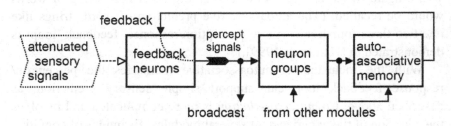

Fig. 13.8. The perception/response feedback loop facilitates the conscious introspection of imagined entities as virtual percepts. If sensory signals are attenuated, the percepts represent the feedback signal patterns in terms of sensory percepts and qualia. In practice, a number of perception/response feedback loops would work together by broadcasting and receiving broadcasts from the other loops.

The reuse of the perception/response feedback loop for imagination allows also the coupling of imagination to actual perception. Imagined objects and actions can be "overlaid" on top of real perceived objects. This is a very useful property; in fact, without this connection imagination would be rather useless. For instance, it is possible to see an object at a certain location and imagine an action that takes this object to another location. This is a simple imagined plan that is directly coupled to the perceived environment.

Imagined plans do not have to be executed at once. The outcomes of the imagined actions may also be imagined and only those imagined actions that would seem to lead to a good outcome, would (or should) be executed. Cognitive prediction and mental rehearsal are also acts of imagination.

Imagination and perception are also connected in another way. Perception delivers only a limited amount of information. It is not possible to see behind objects, but it is possible to imagine what is there. In this case, the act of imagination would be an autoassociative process. Heteroassociative imagination is also possible; this would combine information from several modalities. In Fig. 13.8 this would take place via the broadcast operation and the associative inputs of the associative neuron groups. This would allow, for instance, the imagination of the causes and sources of heard sounds. In this way imagination is related to the looking for meaning.

On a more complex level, imagination is also related to understanding. For instance, we can see what a person does, but we will have to imagine the reason for these acts.

The author has proposed some enabling functions for imagination [Haikonen 2005] and these are:

- The evocation of mental representations of imaginary objects at different positions; what, where
- The evocation of mental representations of change and motion
- The evocation of mental representations of relation: Relative position, relative size, etc. and relative motion; collide, pass by, take, give, etc.
- Mental modification: Make larger, smaller, rotate, combine, move from one position to another, etc.
- Attention, introspection
- Decision making; the selection between competing imagined scenarios.

These functions can be realized by associatively cross-connected perception/response feedback loops with emotional good/bad evaluation and the use of distributed signal representations. Distributed signal representations are suitable for imagination, because they have a fine structure that allows the modification of the imagined entities. Each signal in a distributed signal array represents a feature. By switching off and on these features a depicted entity can be modified to suit the requirements of the imagined situations.

Summary

- Sensory perception begins with transduction; sensors transform the physical phenomena into common internal form.
- Processing with distributed representations call for the detection of elementary features.
- Perception is not a linear stimulus-response process, it utilizes also feedback from the system.
- Feedback facilitates priming, prediction, attention, match, mismatch and novelty detection and also the introspection of mental content as virtual percepts.
- The perception/response feedback loop is a circuit module that realizes the requirements for feedback assisted sensory perception and introspection.

Chapter 14

Examples of Perception/Response Feedback Loops

14.1. The Auditory Perception/Response Feedback Loop

14.1.1. *The Purpose*

The main purpose of the auditory perception/response feedback loop is to create a perceived phenomenal sound scene by segregating individual sounds and externalizing their location. Therefore, the design of an auditory perception/response feedback loop should be based on the special nature of sound, the requirements of direct and transparent perception that conserves amodal qualities and the requirement of the externalization of the sound percepts. Auditory preprocessing is needed to satisfy these requirements.

Additional important functions of the auditory perception/response feedback include the prediction of sounds and sound patterns, instant replay memory and the facilitation of the inner speech via auditory introspection. The prediction of sound patterns is related to the recognition of rhythm. The instant replay memory function corresponds to the *echoic memory* of the human brain. The echoic effect extends the apparent presence of transient sounds and in this way helps to focus attention on these even when these are no longer actually present.

14.1.2. *Auditory Pre-Processes*

The purpose of auditory preprocessing is the production of auditory feature signals that represent the desired information and are suitable for associative processing with distributed signals. Auditory perception is based on the detection of air pressure fluctuations within the audible frequency range, which is usually taken to cover the frequencies from 20 Hz to 20 000 Hz. Sounds with frequencies below 20 Hz are called infrasound and sounds above 20 000 Hz are called ultrasound. Some animals are able to hear and utilize ultrasound frequencies. Ultrasound frequencies have shorter wavelengths and allow therefore more accurate sound source direction detection and echolocation. In artificial systems the detectable audio frequency range may be tailored to suit the requirements of each application.

Auditory pre-processing must facilitate *sound segregation*. The ear or a microphone receives the sum of the air pressure variations generated by every sound that is locally present. A microphone transduces this air pressure fluctuation into an analogical temporally varying voltage, Fig. 14.1.

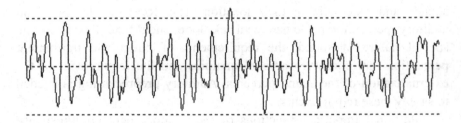

Fig. 14.1. The sound signal from a microphone over a period of time is the cacophonic sum of all incident individual sounds. This signal does not reveal directly the individual sounds.

Figure 14.1 presents the sound signal from a microphone over a period of time. This signal is the cacophonic sum of all incident sounds and as such, it does not directly reveal the individual sounds, there is no segregation. Sound segregation, the separation of individual sounds is a necessary condition for *content detection* and the *association of meaning*. The sum signal does not tell directly much about the content and there is no way, in which meanings could be associated with the separate sounds of this sum sound. Therefore, a preprocess that would allow the extraction of separate sounds is necessary.

As an auditory pre-process, the ear performs a kind of *audio spectrum analysis* and is able to group sounds according to their harmonic contents. The use of two ears allow also the detection of the direction of each sound. Artificial sound perception should utilize similar pre-processing methods.

It is known that mathematically every sound can be considered as the sum of sine waves of various frequencies. A periodic sound consists of the sum of sine waves with a fundamental frequency and a number of harmonic frequencies that are multiples of the fundamental frequency. Consequently, a periodic sound can be extracted from a sum of sounds by determining its fundamental frequency and picking up its harmonics. This can be performed, for instance, by the so-called *comb filtering*. In this way, various sounds can be segregated from each other by the use of frequency analysis.

Spoken language has additional requirements for auditory perception. Heard words must be recognized and learned. This calls for the extraction of audio spectrum, transients and temporal duration. Spoken words are temporally fleeting strings of auditory patterns, called *phonemes* and consequently suitable short-term memories for the capturing and processing of the whole words are required.

The human voice consists of a fundamental frequency and a large number of harmonic frequencies that are multiples of the fundamental frequency. The fundamental voice frequency is also known as the *pitch*, as it is also the distance between the harmonic frequencies. Pitch determines the overall impression of the voice. Typically, the female pitch is around 200–230 Hz and the male pitch is around 110–130 Hz. This explains why female voices sound higher.

The intensity of the harmonic frequency components diminishes towards higher frequencies so that the upper frequency limit of the telephone, 3400 Hz, suffices for communication. For a more natural reproduction of speech the upper limit of 5000 Hz would be preferred and is used in AM radio broadcasts.

We can understand the speech of others regardless of the pitch. In fact, pitch is not very important for the recognition of phonemes, while it may help us to recognize the speaker. Phonemes include vowels (*a, e, i, o, u,* etc.) and consonants (*b, c, d, f, g, h, j, k, l, m, n, p, q, r, s, t, v, w, x, z,* etc.). A spoken vowel is formed, when the spectrum of the harmonic frequencies of the voice is filtered by the mouth and lips. This causes maximums and minimums to the overall spectrum. Mouth cavities are about the same size for every person and therefore these maximums and minimums for each vowel occur at certain rather fixed frequency bands regardless of the speaker, male or female. The frequency bands for maximums are called *formants*. Vowels may therefore be recognized by determining the relative energy within each formant frequency band and these quantities would be suitable auditory features for vowels.

Consonants are characterized by transients, the abrupt changes in sound intensity. The spectrum of a transient sound is rather continuous towards the high end of the frequency range and does not consist of a fundamental frequency and its harmonics.

Distributed signal representations call for the detection of elementary features. According to the above, useful auditory elementary features may consist of the energies within narrow frequency bands that would cover the whole auditory frequency range. This calls for audio spectrum analysis that resolves narrow frequency bands and their energies. There are two methods that could be used; these methods are the *Fourier transform* and the use of *filter banks*. The ear uses a method that is effectively a filter bank.

Sounds and pronounced phonemes may have different durations, therefore the duration of sounds and phonemes must also be determined and represented as a distributed signal array. The detection and representation of sound duration would also allow the representation of rhythm.

14.1.3. *The Outline*

As an example, a simple auditory perception/response feedback loop that includes some of the basic functions is outlined here, see Fig. 14.2.

This auditory perception/response feedback loop utilizes a filter bank for the audio frequency analysis. The filter bank consists of a number of narrow bandwidth filters. The output signals of these filters describe the instantaneous energy of each narrow frequency band and constitute the auditory feature signals. These signals are forwarded to the feedback neuron group. The output signals from the feedback neuron group constitute the serial stream of auditory percepts that is broadcast to the other modules of the cognitive system. This serial broadcast preserves amodal rhythm. This property is necessary for any synchronized motor actions including the production of speech.

There are two parallel associative neuron group assemblies within the feedback loop. Both these assemblies learn and reproduce also temporal durations and rhythm. The serial autoassociative memory allows the instant replay of the heard sound as well as the prediction of a sound sequence. The sequence evocation neuron group allows the association of sound sequences with cues and the evocation of these sequences with the same cues. The winner-takes-all (WTA) circuit selects the relevant output.

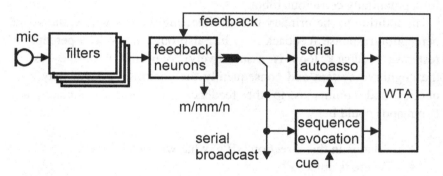

Fig. 14.2. A simple auditory perception/response feedback loop. The sound is analyzed with a filter bank. The intensity values from the filters constitute the auditory feature signals. The serial autoassociative memory allows the instant replay of the heard sound as well as the prediction of a sound sequence. The sequence evocation neuron group allows the association of sound sequences with cues and the evocation of these sequences with the same cues. The winner-takes-all (WTA) circuit selects the relevant output. Serial broadcast preserves amodal rhythm.

During listening to music the feedback tracks and predicts the music in real time. The autoassociative memory predicts the continuation, which will coincide with the actual perceived continuation. It is also possible to focus attention on a specific instrument by the priming effect of the feedback (for priming see Fig. 13.3).

14.2. The Visual Perception/Response Feedback Loop

14.2.1. *The Purpose*

The main purpose of the visual perception/response feedback loop is to tell the cognitive system what is where. In a potentially conscious robot the visual perception/response feedback loop should be able to create the impression of a seen, inspectable external world and the visual percepts should appear as externalized visual qualia. All visual qualia in a robot, such as colors, do not have to be similar to the human visual qualia, but they should nevertheless appear as properties of the seen world, instead of some properties of the internal electronic activity. This is the main requirement for the design of a visual perception/response feedback loop for a potentially conscious robot.

In addition to the primary task of detecting what is where, the visual perception/response feedback loop has also other tasks. The perception/response feedback loop is associatively connected with the rest of the cognitive system and consequently the cognitive system may send queries and commands to the feedback loop. Typical queries and commands could be:

- What is in this direction? (E.g. what was behind me?)
- Where is object x?
- Is this object x?
- Find x!
- Imagine this kind of object!

The visual perception/response feedback loop could broadcast responses like these:

- This is what is here.
- The object x is here.
- The object x has been found.
- This object is or is not the object x.
- The object is moving this way.
- Unexpected objects or action has been detected.

Figure 14.3 illustrates the tasks of the visual perception/response feedback loop.

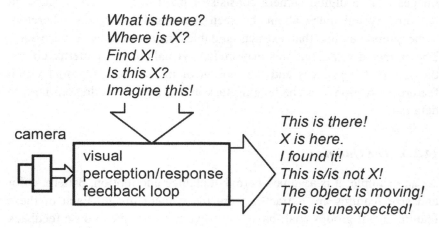

Fig. 14.3. Tasks for the visual perception/response feedback loop.

The successful execution of these tasks calls for the ability to detect objects, the direction where they are and the short-term memorization of these objects and their locations. In addition, the perception/response feedback loop must be able to detect match/mismatch and novelty conditions. The visual perception/response feedback loop is also the enabling machinery for visual imagination and the introspection of the imagined entities.

14.2.2. *Visual Pre-Processes*

The purpose of visual preprocessing is the production of visual feature signals that represent the desired information and are suitable for associative processing with distributed signals.

Visual perception is based on the detection of the light intensity and wavelength that is reflected by the observed objects. Advanced visual perception utilizes imaging systems such as the eye or a digital camera, which are able to create optical projections of the environment on a photosensitive surface. However, simpler visual perception systems also exist and are useful.

The visual perception/response feedback loop module shall detect and broadcast visual features and the direction, where they appear. The images from a digital camera consists of pixel maps, which as such do not directly tell much about the seen scene and its objects. Therefore some preprocessing that extracts suitable visual features is necessary. The extracted visual features may include visual change, patterns, colors, the size of the patterns and the motion of the patterns. Optimal visual feature extraction is a challenging task and cannot be addressed here in detail.

14.2.3. *The Outline*

As an example, it is assumed here that change and motion, pattern, color and size features are extracted from the camera image. Each of these feature signal groups shall have their own perception/response feedback loop and the complete visual perception/response module consists of the interconnected visual feature perception/response feedback loops, see Fig. 14.4.

In Fig. 14.4 it is assumed that the image sensor of the camera has a small high resolution center area, which determines the focus of visual sensory attention. The direction of the optical axis of the camera is then the gaze direction towards the object. A well-defined gaze direction allows the pinpointing of objects. The gaze direction can be changed by turning the camera in pan and tilt directions.

The pan and tilt motors are controlled by the direction perception/response feedback loop, which senses and broadcasts the instantaneous direction of the camera. When the camera pans and scans the visual scene, the instantaneous direction is associated with the instantaneously perceived visual features (connections S2, C2, P2, M2) and in this way short-term memories of the scene are created.

Short-term memory imagery, such as what was previously seen in certain direction, may be evoked by the associative evocation of a direction (connection D1) by an externally applied signal pattern. The evoked direction is returned into a direction broadcast via the feedback loop and this direction is then broadcast to the other visual feature feedback loops (connections S2, C2, P2, M2). There the direction evokes the visual features that were associated earlier with that direction. The evoked visual features are returned into virtual percepts via the feedback loops. In this way questions like "what was behind me" can be answered.

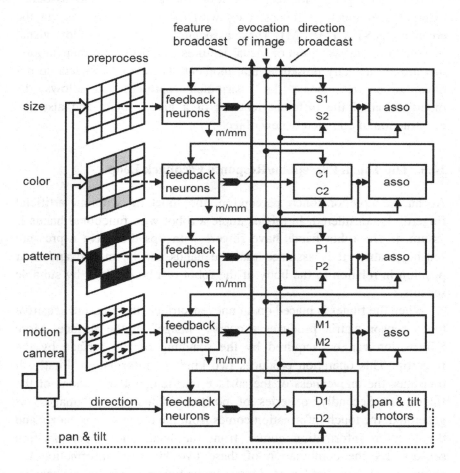

Fig. 14.4. The visual perception/feedback loop module consists of a number of sub-modules, such as the pattern, color and motion modules. Each of these have their own feedback loops.

The change and motion perception/response feedback loop is also connected to the direction perception/response feedback loop. With the provided visual change information the pan and tilt motors are able to turn the camera towards the visual change. This function allows also the tracking of a moving object.

The visual perception/feedback loop module is able to associate external cues (such as names, etc.) with the percept features via the connections S1, C1, P1, M1. These associative connections allow visual imagination, the evocation of visual features of objects and their motion that are not actually present at that moment. The evoked objects do not have to be real ones; the distributed representation allows the modification of the different features of a given object and also the construction of completely new objects.

14.3. The Touch Perception/Response Feedback Loop

As an example of touch perception, the touching with an artificial fingertip is considered. In this example a robot with functional hands is assumed. The robot hands have fingers with tips that have a pressure sensing matrix. It is assumed that the hands can be moved and their position in relation to the body of the robot can be detected by suitable sensors.

When the finger is placed on an uneven surface, the pressure sensitive touch sensor matrix produces a coarse "image" of the surface. More information can be acquired by the scanning of the surface by the fingertip. This operation produces two kinds of information. Firstly, it translates the unevenness of the surface into temporal vibrations of the finger tip. Secondly, a series of position-related touch "images" is generated; the touch information comes from the touch sensor matrix and the position information comes from the hand and finger position sensors. By the combination of these two kinds of information the contours and shapes of larger objects can be felt and their "image" can be constructed in a memory.

The touch sensor matrix does not broadcast any position or location information to the cognitive process. The perception of the location of the touch percepts as the point of the finger tip arises from the addition of the position information of the hand and fingers. This information can originate from the position sensors and can also be achieved visually, if the robot is sighted. With these means the touch percepts are externalized to the tips of fingers and a contribution to the body image is made.

According to the aforesaid, touch perception involves the cooperation of at least two sensory modalities, namely the touch modality and the body position (proprioception) modality. This calls for the interconnection of the related perception/response feedback loops, Fig. 14.5.

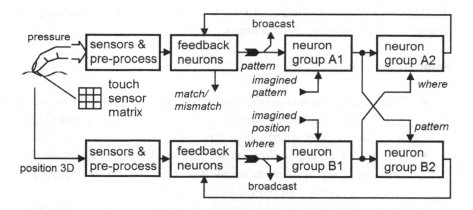

Fig. 14.5. The perception/response feedback loops for touch perception.

Figure 14.5 depicts the interconnected touch and position perception/ response feedback loops for touch sensing.

The touch sensor matrix delivers pressure-related signals from each individual pressure sensor to the feedback neurons. The group of these signals is a kind of a coarse image that represents the structure of the sensed surface.

The position perception/response feedback loop delivers the corresponding location of the fingertip. The position of the fingertip is sensed by a variety of proprioceptive sensors that detect the angles of each joint in the hand-finger mechanical system. The resulting percepts define the position of the fingertip in three dimensions.

The neuron groups A2 and B2 are associative short-term memories that cross-associate the touch image and its location with each other. The neuron groups A1 and B1 are non-volatile associative memories. The neuron group A1 associates a name or some other cue with the touch pattern and the neuron group B1 associates similarly a cue with a corresponding position. Thus, a position cue at the neuron group B1 will evoke the corresponding position neural pattern. This position information, "where", is forwarded to the neuron group A2, where it will evoke the latest touch percept at that position. When an object is felt by scanning its contours by the finger, more or less complete information about the shape of the object is accumulated in the neuron groups A2 and B2. This information will allow, for instance, the finding and touching of a certain point of an object in darkness.

Summary

- The auditory perception/response feedback loop should create a perceived phenomenal sound scene by segregating individual sounds and externalizing their location.
- The visual perception/response feedback loop should create the impression of a seen external world and the visual percepts should appear as externalized visual qualia.
- Touch perception should combine touch pressure sensations with the hand and finger position sensations so that an externalized "image" of the touched object could be created.

Chapter 15

The Transition to Symbolic Processing

15.1. From Distributed Signals to Symbols

A perception/response feedback loop operates inherently with sub-symbolic representations, arrays of distributed signals. These signals indicate directly presence of the corresponding sensorily perceived or introspected features, but apart from that they do not have other inherent meanings. These kinds of direct sub-symbolic representations are a necessary prerequisite for possible qualia and must therefore be utilized in potentially conscious machines. However, higher cognition calls for symbolic presentations, too. The use of natural languages is based on the symbolic use of sound patterns (words) and visual patterns (text). Abstract thinking and reasoning are also based on symbolic processing. Symbols depict entities that are not inherently related to them and are beyond the direct meaning of their constituting features; symbols get their meaning via convention. Therefore, in order to use percepts as symbols, additional meanings must be associated with them.

15.2. Requirements for a Natural Language

A robot can be conscious without the mastery of a natural language as long as the requirement of the inner appearance of the subjective experience is fulfilled. However, a proposed artificial cognitive architecture cannot be considered to be completely plausible, if it is not able to support even a most rudimentary natural language at all. This is a philosophical issue, if the proposed cognitive architecture is supposed to be a model for human cognition. This is also a practical issue; successful cognitive robots should be able to converse with humans using a natural language.

Superficially, the use of language might not seem to be a big issue. After all, we already have artificial systems that communicate with their human users using a natural language. Elevators talk and computers display messages using English or some other language. However, there is a difference that should be noted. The existing systems that use natural language do not have inner speech and they do not really understand, what the communication is about. Their natural language communication is only a mechanical preprogrammed routine.

True conscious robots should be able to learn and understand a natural language and utilize it in a meaningful way, not as some preprogrammed strings of words. This leads to system requirements that are discussed in the following.

As stated before, a natural language is a symbol system that allows the description of situations by strings of words and sentences. A natural language has a vocabulary and syntax. The vocabulary is the collection of the available words and the syntax gives the rules for the indication of those conditions and relationships that cannot be conveyed by the basic words alone.

The vocabulary of a cognitive agent is not only a list of available words. The words are symbols with meanings, which are conveyed by associative connections. The meanings of the words are grounded to the perceived entities and situations of the external world and the body and also to more abstract concepts. This leads to the requirement of a perceiving system with associative learning and associative processing of information.

Words with pointable simple meanings are learned first. This is a naming process (also known as ostension); an object, property or action is pointed out and named. In this process the sub-symbolic signal patterns of the pointed entities and those of the given word are associated with each other, so that afterwards the signal patterns of the word can act as a symbol for the named entity. This process calls for a system that is able to perform the transition from sub-symbolic to symbolic information processing.

15.3. Association of Meaning

Percepts may have associated meanings. A tool is not only a perceived object, its percept evokes also possibilities for its use. A perceived entity may also be treated as a symbol. In that case it can be used cognitively instead of the percepts of the entities that it represents. These kinds of associated meanings call for neural mechanisms that allow the association of percepts from various modules with each other. As an example the association of a spoken word with a visual pattern is presented. This operation calls for associative cross-connections between the visual and auditory modules, see Fig. 15.1.

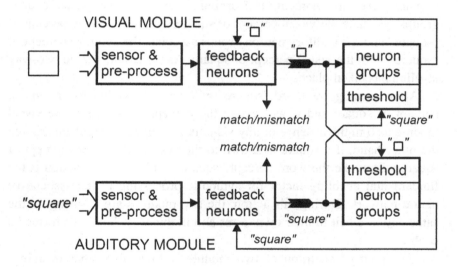

Fig. 15.1. Associative cross-coupling of modules allow the association of meaning. Here two modules, visual and auditory, are associatively cross-connected. In this example the word "square" is associated with the pattern "□". When the visual module detects a "□", the name "square" is evoked in the auditory module and vice versa. (The operation of the neuron group is explained in chapter 12.)

In Fig. 15.1 the visual module has an image sensor that is able to capture images of external objects. The image information is preprocessed and a number of feature signals are forwarded to the feedback neuron groups. The auditory module is able to detect spoken words, which are processed into a number of feature signals. Without associative cross-connections these words would be without meaning.

The cross-connection lines between the two modules enable the formation of associative cross-connections. These connections are learned and are activated by the presence of an auditory or visual percept that is associated with a percept at the other modality. At other times the cross-connection between the modules is not active.

During associative learning, the percept from the visual module is forwarded to the associative input of the auditory module neuron groups and vice versa. Let us assume that the visual module is perceiving a square and at the same time the auditory module is perceiving the word "square". At this moment, if learning is enabled, the visual neuron groups associate the visual percept of square with the auditory percept of "square" and the auditory neuron groups associate the auditory percept of "square" with the visual percept of square. As the result, simple two-way labeling has taken place.

After learning, the word "square" will act as a symbol for the visual percept of square and it will evoke the percept of square at the visual module also in the absence of any visually perceived external square. On the other hand, if a square is shown to the system, the visual percept of square will evoke the word percept "square". This two-way action is the fundamental enabling factor for symbolic processing in the associative neural network. During this action, the two modules are focusing on the same entity, albeit in their own terms and in this way they have formed a coalition.

This kind of coalition of two modules allows also other meaning-related functions. For instance, if in the previous example the word "square" and the actual square were presented at the same time, match-conditions would occur at the feedback neurons. Thus, with the association of the words "yes" and "no" with the match and mismatch conditions, the system will be able to answer "yes" to the question: "Is this a square?". If the presented word and the visual object do not match, then mismatch conditions occur and the system will answer "no" [Haikonen 1999]. It is obvious that this simple two-way labeling and naming allows also rudimentary verbal communication.

Chapter 16

Information Integration with Multiple Modules

16.1. Cooperation and Interaction of Multiple Modules

Useful cognition calls for the utilization of meaning and understanding and the evoking of affordances that would allow meaningful action. These requirements go beyond the possibilities provided by the simple two-way labeling and consequently call for information integration and the interaction between several modules.

As an example of the associative cross-connection and interaction between multiple modules, an extension of the previous examples is presented. In this example the cooperation and associative interaction of four modules is described; the modules are the linguistic auditory module, visual pattern module, the visual color module and the motor module. The relevant associative interconnections of these modules are given in Fig. 16.1.

In this example the process begins, when the auditory module receives the verbal request *"find cherry"*. These words are broadcast to the visual and motor modules. The word *"find"* is associated with searching at the motor module and accordingly it will trigger a search routine, such as the turning of the head. The motor module does not have to know, what is being searched for and consequently the word *"cherry"* is not associated with any motor action and is ignored. The visual modules ignore the word *"find"*, because it is not associated with any visual object. The word *"cherry"* is associated with a pattern and a color and will evoke the pattern feature *<round>* at the visual pattern module and the color feature *<red>* at the visual color module. These are fed back to the corresponding feedback neuron groups P and C.

When a cherry is found and seen, the sensory features of *<round>* and *<red>* match the feedback features and match signals are generated. A *match*-signal is also broadcast to the motor modality, which now terminates the search operation leaving the perceived object in the focus of visual attention.

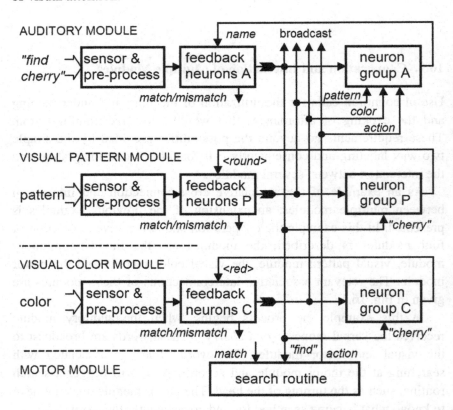

Fig. 16.1. The associative cross-connection of four modules. When the auditory module receives the request *"find cherry"*, the word "cherry" evokes the virtual features *<round>* and *<red>*, which are compared with the pattern and color percepts. When the evoked features match the perceived features, the object has been found and the match-signal terminates the search action.

The pattern and color percepts are also broadcast to the auditory module, where the names of the pattern and color may be evoked, but also their combinations may evoke the name of the corresponding object, if it has been previously named. In this case the word "cherry" might be evoked. The *action* signal pattern from the motor module is broadcast to the auditory module and this interconnection allows the verbal naming and reporting of the on-going motor activity.

The network of associative cross-connections in this example can be easily extended to systems with a larger number of modules based on the perception/response feedback loop principle.

16.2. Sensorimotor Integration

It was argued earlier that the conscious perception of the environment involves the externalization of the internal phenomenal appearance of the related neural activity. Due to this externalization the world appears to be out there, outside our brain and body. This outside-the-body externalization effect is present with the non-contact senses of vision and hearing. (Externalization applies also to body sensations making them to appear outside the brain.)

The externalization of the appearance of sensory information allows *seamless sensorimotor integration*, the direct interaction with the environment, because the externalization adds the sense of direction and distance to the perceived object. When we see a book on a table, we can reach out and pick it up without any conscious symbolic computations for the necessary hand movements. The perception of the external position of objects within our reach is seamlessly coupled to the neural circuits that control the motion of our hands and body, allowing thus the possibility for direct motor actions. The readiness for these actions arise automatically as soon as objects are perceived, without any special preparations; the command to execute the act would arise from the cognitive context. This seamless connection between the senses and motor systems (muscle control) is called sensorimotor integration.

Traditional robots do not have this kind of seamless sensorimotor integration. There is no subjective experience with an internal appearance at all and consequently no externalization effects can take place; there is nothing to be externalized. The digitalization and symbolic processing of sensory information effectively voids this possibility and the robot is reduced to a blind mechanism. Every motion that a traditional robot makes, must be computed by preprogrammed rules. If a robot were to pick up a book, it would first have to determine the exact position of the book and next it would have to compute a suitable trajectory for its hand. During the execution of this task the robot would have to determine the exact position of its hand and compare this with the computed trajectory. Any error in the instantaneous hand position would have to be corrected. True, this can be easily executed by digital signal processing and well-known feedback control loops, but nevertheless, preprogrammed algorithms for each executable motor routine, action and action sequence would be necessary. Computers do not have any "ex tempore" action modes and cannot improvise any new action routines on the fly.

Associative information processing offers another, direct way for the realization of sensory percept externalization and sensorimotor integration, without any computations or preprogrammed motor routines. However, associative robots would still use conventional electric motors, which do not have associative interfaces. Therefore, the practical design of associative sensorimotor integration calls for the design of interfaces that allow the control of motors by associative distributed signal representations. In the following, the traditional motor feedback control loop is explained first and then its connection to an associative perception/response feedback loop is described. Next, it is explained how this arrangement enables seamless sensorimotor integration.

16.3. Feedback Control Loops

A typical effector system consists of a motor that moves some moving part, for instance, a robot hand. The instantaneous position of this moving part is detected by a position sensor that outputs a signal that is proportional to the sensed position. The system is controlled by an input signal that indicates the desired position for the moving part. The difference between the desired position (set-value) and the sensed position (is-value) is computed (in analog systems no numeric computations take place, the difference can be determined by simple resistor circuits). If the difference between the sensed position and the desired position is zero, then the moving part is at the desired position. If the difference is not zero, then the polarity of the difference signal indicates the direction that the motor must run so that the moving part can achieve the desired position. This system arrangement is called a *feedback control loop*, Fig. 16.2.

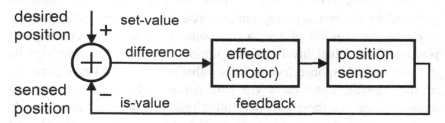

Fig. 16.2. The principle of the feedback control loop for a motor (effector control loop). The effector moves a moving part whose exact position is measured continuously by the position sensor. The difference of the desired position (set-value) and the measured position (is-value) is continuously computed and this difference drives the effector. When the difference is zero, the desired position has been reached and the effector stops.

Figure 16.2 depicts a simple proportional feedback controller. This arrangement works, but is not very accurate. Obviously, when the desired position and the sensed position are very close, the difference signal will become very small and the motor may no longer have a sufficient drive signal. This leads to a small steady-state position error for the moving part. In industrial applications, proportional-integral-derivative controllers (PID controller) are often used, because their steady-state error is zero. For the purposes of this discussion, the proportional feedback controller model is sufficient.

In order to make a feedback controlled effector to execute a desired motion, two signals must be provided, namely the desired position signal (set value) and the signal that indicates the actual sensed position (is-value). In analog systems continuous signals (voltages) are usually used; the amplitude of these signals is proportional to the position values.

16.4. Hierarchical Control Loops

The cognitive control of motor actions is hierarchical. The actual desire to execute a motor action, such as the grabbing of an object arises at circuits that are not directly related to the actual motor circuitry. A seen object may evoke the desire to pick it up; this could be a conscious general level command that would have to be transformed into detailed (but sub-conscious) set-values that can be accepted by the actual motor control loops, such as that of Fig. 16.2.

The desired position that a robot hand should reach, could be provided by the visual perception/response feedback loop. This position might be the position of a seen real object or it might be an imagined position. The actual hand position information would be provided by the perception/response feedback loop that senses the relative positions of the moving body parts (*proprioception*). However, the position representations of these two perception/response feedback loops would not be compatible with each other. The visual position percepts and the proprioception percepts would be represented in terms of their corresponding sensors and therefore the resulting signal patterns would not have one-to-one signal-wise correspondence. Therefore the difference between the sensed position and the desired position cannot be determined directly from these signal patterns. This compatibility problem can be overcome via hierarchical associative connections.

Figure 16.3 depicts the hierarchical control mode of action that is effected by the cross-connection of the visual perception/response feedback loop and the proprioception perception/response feedback loop. The desired visual position is broadcast to the associative neuron group at the proprioception loop, where it evokes the corresponding position representation in proprioception terms. The actual motor action is executed by the effector loop, which is similar to that of Fig. 16.2. The effector loop will run the motors until the proprioceptive is-value matches the set-value.

In Fig. 16.3 proprioceptive sensors detect the current position of the moving part and forward this in information in the form of distributed signal representation to the feedback neurons. Normally, this representation would also be the percept representation, which would be broadcast and also forwarded to the inner associative neuron group. The output of the associative neuron group would be the set-value and the output from the proprioceptive sensor would be the is-value for the effector loop.

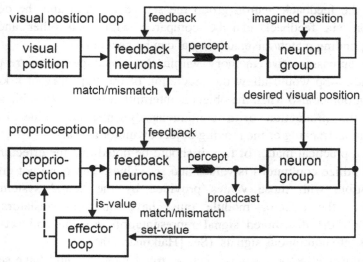

Fig. 16.3. The associative motor control system based on a perception/response feedback loop. Proprioception gives the current position of each body part (is-value). The desired visual position evokes the corresponding set-value for the effector. The effector loop has built-in translators that transform the distributed signal representation is-value and set-value signals into continuous signals so that their difference can be determined. The proprioception feedback loop allows also the virtual execution of motor actions.

Both the desired position and the actual position information would be in the form of distributed signals. The proprioception and visual perception systems would be separate modalities and therefore their distributed signal representations would depict dissimilar quantities; there would not be any common code. Therefore the desired position distributed signal representation must be translated into the quantities that are used by the proprioception loop. This can be done by neuron groups via associative learning. During an initial learning phase the system of Fig. 16.3 would visually perceive various positions of the moving part and these position representations would be associated with the corresponding proprioceptive representations at proprioception loop.

Thereafter a visual position percept would be able to evoke the corresponding proprioceptive representation. It should be noted that the desired position may not be necessarily seen, it may also be imagined (one may close eyes and reach out towards an imagined position).

At the feedback neuron group the sensed position and the desired position via feedback can be compared with each other and the corresponding match/mismatch signal can be generated.

An autoassociative memory within the visual perception/response feedback loop would allow the prediction of the trajectory of a known moving object also when the object is intermittently not seen. This would allow the production of a continuous dynamic set value for the continuous tracking of the moving object by motor actions.

As a practical matter of fact, it should be noted that the effector loop would utilize continuous is-values and set-values. These are not directly compatible with those values provided by the neural system and therefore the effector module must have built-in translators that transform the distributed signal representation is-value and set-value signals into continuous signals. (See [Haikonen 2007]).

Practical effector systems, such as robot hands, would have several joints and motors. Every joint would have position sensors and the motors would have an associative control loop.

The function of the above associative neural motor control enables the seamless integration of the visual modality and the motor modality, the sensorimotor integration. The sensorimotor integration is one enabling factor for the externalization of non-contact percepts; reaching out for an object means that the object is out there.

Summary

- Cognition and meaningful action call for associative cross-connections and interaction between multiple modules.
- For the execution of motor actions feedback control loops are usually required.
- Sensorimotor integration allows a robot to combine visual, auditory and other sensory information with the motor systems allowing effortless operation in all environments.

Chapter 17

Emotional Significance of Percepts

17.1. The Significance of Percepts

In principle, an associative information processing system is able to accumulate a large number of possible associative connections between entities. Any stimulus would activate many associative links and as a result many signal patterns would be evoked. This leads to the problem of choice, because the evocation of all possible associations would overflow the system. This serious issue is related to the so-called *frame problem* and *combinatorial explosion*. Therefore the space of choice must be limited in one way or another. In associative information processing, the associative evocation process must be guided in such a way that at each moment only the most relevant associative connections would be activated. This calls for mechanisms that are able to guide attention and pinpoint the most important associations. Context is one such mechanism that is easily realizable in associative systems, but it is not sufficient in every situation.

An example along the classical one provided by Dennett [1987] illustrates the problem of choice: A robot enters a room, in order to find and retrieve a certain object from a large number of various objects. Among these objects is a bomb and its fuse is burning. What should the robot do? Should the robot go there and start looking for the object to be retrieved? This might take a while and in the mean time the bomb would explode. Obviously the robot would have to focus its attention on the bomb immediately. However, how could that happen, if the robot were guided by context only, which in this case would be related only to the object to be retrieved? Generally, context would frame and limit the

163

scope of choice and should be used as an attention guiding factor, but it does not really help in situations like this. A context-guided robot would be looking for a certain object and while doing so, it would ignore all the other items that do not fit the context and do not resemble the searched object. In doing so, it would also ignore the bomb, because in the robot's mind the bomb would have the same irrelevance as all the other unrelated items. Ignoring the bomb would be a stupid thing to do, but there is nothing in the context that would alert the robot about the immediate danger.

Obviously, in addition to the context, there is a need for a mechanism that would be able to evaluate instantly the significance of each perceived object and situation. If necessary, this evaluation should be able to override the context-related attention and focus attention on the more significant percepts.

17.2. Emotional Evaluation of Percepts

Emotional value systems are mechanisms for the evaluation of the importance and significance of percepts. These systems learn and associate emotional significance values with percepts and use these values to focus attention on most important percepts. These systems can also work as motivational drivers.

A simple emotional value system utilizes pleasure and pain sensors and additional neuron groups that are able to associate pain and pleasure signals with various percepts from other sensory systems. The purpose of this arrangement is to provide pleasure and pain signals for the evocation of various system reactions and the amplification of the broadcast percept that has strong emotional significance value. A simple emotional value system is depicted in Fig. 17.1.

The emotional value system of the Fig. 17.1 consists of pleasure and pain sensors and pleasure and pain neuron groups. This assembly is connected to a sensory perception/response feedback loop. (In actual system realizations there would be several perception/response feedback loops instead of the one that is depicted in Fig. 17.1, but the general principle would be the same.)

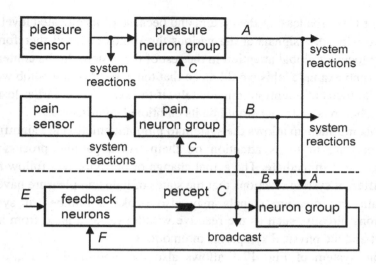

Fig. 17.1. A simple emotional value system and its connection to a perception/response feedback loop. Percepts are associated with simultaneously occurring pain or pleasure and later on will evoke the associated pain or pleasure signal. This will focus attention on the percept.

The system of Fig. 17.1 associates a sensory percept C with pleasure A or pain B, if these occur simultaneously with the percept C of an event E. These associations take place at the pleasure and pain neuron groups that receive the percept broadcast C as their associative input. The active pleasure and pain signals A and B are also present as associative inputs at the perception/response feedback loop neuron group. Here another association between pleasure A or pain B and the sensory percept C may take place.

Let's assume that pain B has been associated with the percept C. In this case the next occurrence of the sensory percept C will evoke the associated pain signal B. This evoked signal is forwarded to the neuron group at the perception/response feedback loop. Here this signal B evokes the same percept signal C internally. This evocation is returned to the feedback neurons as the feedback signal F via the feedback loop and consequently the percept C will be amplified. The intensity of the percept signal carries the emotional significance of the percept and thus the emotional significance of the signal C has been elevated. The amplified percept C is broadcast to the system and because of its elevated level it is able to win other signals at the various thresholds and will therefore be on the focus of global attention in the rest of the system. In the context of the "bomb example" this would mean that the percept of the bomb would gain the focus of attention immediately, if the system had earlier learned what a bomb is and what would be its emotional significance.

This mechanism allows the system to pinpoint emotionally significant percepts and to focus attention on them. As a parallel process this evaluation is immediate. The actual change of behavior will follow from the different system reactions that the states of pain and pleasure have.

Pain and pleasure signals may also evoke a number of system reactions directly, such as the reactive withdrawal of a hand from a hot object and the physical symptoms of emotions.

The system of Fig. 17.1 allows also the emotional evaluation of internally evoked percepts, such as the products of imagination. This evaluation could also lead to the evocation of system reactions, perhaps weaker ones. Direct system reactions and those evoked by emotional significance evaluation are rather important also to simple associative robots, while more complicated emotional states may not be.

Summary

- Combinatorial explosion problem haunts systems with large number of possible interactions and combinations.
- Frame problem relates to efforts to limit the combinatorial explosion by limiting the scope of choice by context-related criteria.
- Emotional evaluation and significance can be used to augment the context-based framing of the scope of choice.

Chapter 18

The Outline of the Haikonen
Cognitive Architecture (HCA)

18.1. General Overview

The Haikonen Cognitive Architecture (HCA) [Haikonen 2003, 2007] combines the afore presented principles and circuit solutions into a dynamic system for the production of possibly conscious human-like cognition including the flow of inner imagery and natural language inner speech.

As previously noted, consciousness is based on perception with qualia and therefore every possibly conscious machine should incorporate qualia-based perception as an essential function. Therefore also the HCA is designed around a sub-symbolic perceptual system that allows percept externalization with the help of seamless multisensory and sensorimotor information integration.

The HCA incorporates associative information processing with distributed signal representations and utilizes emotional significance evaluation and match/mismatch/novelty detection as attention and relevance guiding factors. Information processing is executed by associative neural networks that allow the transition from sub-symbolic to symbolic processing according to the principles presented in the earlier chapters of this book.

The operation of the HCA can be simulated by computer programs, but in this case the sub-symbolic function is lost, the perception process will not be direct and no qualia or perceptual externalization effects will be present.

The general principle of the information flow in the HCA is summarized in Fig. 18.1.

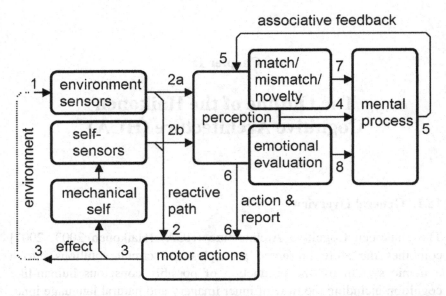

Fig. 18.1. The general principle of the information flow in the Haikonen Cognitive Architecture (HCA). The operation is based on perception process, which allows also the introspection of the results of mental processes as virtual percepts via the associative feedback. Note; speech is also a motor act.

Figure 18.1 depicts the generalized units and the information flow paths between them in the HCA. The environment and the mechanical self (the body and its functions) are sensed by a number of sensors. These sensors output their signals to the perception process and also directly to the motor action unit. The perception process is coupled to emotional evaluation and match/mismatch/novelty detection. They all output their signals and effects to the mental process. The output of the mental process is fed back to the perception process, so that the requirements of priming, prediction and introspection are fulfilled.

The general configuration of Fig. 18.1 allows the following modes of operation:

1. *Reflex reaction.* Stimulus → reflex (1 → 2 → 3). Here the stimuli 2a, or 2b evoke directly motor responses. This corresponds to the stimulus-response reflex, for instance, the

withdrawal of the hand from a flame. This mode of operation does not produce conscious reports of what is happening; reports may be generated afterwards, when the perception process and global attention becomes involved.

2. *Sub-conscious routines.* Stimulus → percept → action (1 → 2 → 6 → 3). This corresponds, for instance, to the automatic manipulation of a small object just because it is within reach, without any real need or global attention focus. This action can be reported if global attention is focused on it.

3. *Deliberated actions.* Stimulus → evaluated percept → associative deliberation → evaluated introspective percept → action (1 → 2 → 4,7,8 → 5 → 6 → 3). This signal path corresponds to a deliberated loop; the motor response arises as the result of perception, evaluation and planning. This action can be reported.

4. *Imagination.* Evaluated introspective percept → associative deliberation → evaluated introspective percept → associative deliberation → ... (4,7,8 → 5 → 4,7,8 → 5 →...). This loop would correspond to imagination and free-running thoughts, where the products of the mental process are perceived and emotionally evaluated, as if they were sensory percepts. These virtual percepts may then be used as the cues for the subsequent thoughts. The results of this process can be reported.

These basic modes of operation of the HCA satisfy the requirements of cognition on a general level. In the following it is described, how these functions are realized with an architecture consisting of a number of cross-connected perception/response feedback loop modules.

18.2. The Block Diagram of HCA

The complete HCA consists of a number of associatively cross-connected modules that utilize the perception/response feedback loop principle. A simplified block diagram depicting the organization of the complete HCA is given in Fig. 18.2.

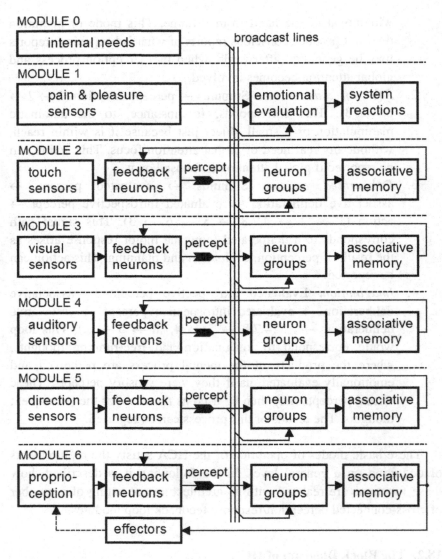

Fig. 18.2. The Haikonen cognitive architecture (HCA) consists of perception/response feedback loop modules that broadcast their percepts to each other. Each module consists of a large number of parallel single signal feedback loops. Each percept consists of an array of feature signals. Interconnecting lines depict paths for signal arrays. Apart from sensors and effectors the blocks depict associative neuron groups.

Figure 18.2 depicts the HCA with seven modules. The number of modules is not fixed to the seven shown here, depending on the application more or fewer modules may be included. The HCA block diagram not a flowchart for a computer program, instead it depicts a dynamic integrated system capable of various parallel and temporal processes and reactions.

Sensory information is preprocessed into feature signal arrays and each module has a dedicated perception/response feedback loop for each signal. Thus, each module consists of a large number of parallel single signal feedback loops. Each percept consists of an array of feature signals. Apart from sensors and effectors the blocks depict associative neuron groups.

In Fig. 18.2 the module 0 consists of internal need detectors such as energy levels etc. (In humans: hunger, thirst, tiredness, etc.). This module generates basic motives for action. The module 1 contains pain and pleasure sensors and acts as emotional evaluator and motivator. The modules 2, 3 and 4 are sensory modules for touch, vision and sound perception. The module 5 detects and keeps track of the direction of body by the help of accelerometers. This information helps the visual module to accumulate and utilize kinds of maps of the environment. The module 6 detects the body part positions and is related to motor effectors and their control circuits.

The HCA is based on sub-symbolic neural circuits. However, the associative processing allows also the association of arbitrary meanings to percepts and the use of these as symbols for the associated entities. This transition to symbolic processing facilitates natural language in the HCA, because all modules broadcast to the auditory module, which is then able to associate these broadcast percepts with auditory neural patterns. Consequently, the percepts of all sensory and motor modalities can be represented by auditory signal patterns, which in this way become linguistic symbols, words, for the represented entities. By broadcasting the words back to the other modules the auditory module is able to evoke kinds of virtual percepts of the described entities and relationships. Moreover, the flow of words can become detached from the actual ongoing activities of the other modalities and can act as a running self-commentary and a vehicle for imagination in the form of *inner speech*. A natural language is, of course, also a means of communication.

18.3. Control, Motivation and Drivers

The HCA is an autonomous system that is driven by internal and external factors. In robotic applications the internal motivating factors may include the need for energy and safe operating conditions. Reflex reactions that support these may be built in. These reactions might include the withdrawal from harmful situations, such as the avoidance of open flame, etc.

The emotional module of the HCA associates emotional values with the percepts of objects and actions and initiates corresponding hard-wired system reactions. The hard-wired system reactions of pain and pleasure are important motivators. Actions associated with pain would be avoided and actions associated with pleasure would be sought for. This mechanism allows the teaching and motivation of some tasks and behavior by reward and punishment; the association of pain and pleasure.

External factors, opportunities, provided by the environment may also motivate and trigger action. An observed opportunity to execute an action would initiate imagined execution of this action. This imagined action would be transformed into real action if no counteracting factors were present. These counteracting factors would include the products of emotional evaluation of the imagined action and any external situation-related factors.

External opportunities as action triggers could also be allowed to lead to the playing and toying with objects. This would facilitate the self-learning of the ways of the external world.

18.4. Information Integration, Coalitions and Consciousness

Cognition calls for multisensory and sensorimotor information integration. Percepts from different sensory modalities must be able to form a coherent view of the world and this view must allow correct motor actions. The percepts must be able to evoke meanings and affordances and their emotional significance must be evaluated. Sensorimotor integration is also required for the externalization of percepts, which would allow the creation of the impression of the observable world out there and also the acquisition of the body image. All this calls for seamless cooperation and coalition of the modules and associative memories. This kind of coalition results in the activation of large number of associative paths. This in turn facilitates the forming of associative memories and the possibility to report the episode later on.

Reportability is one hallmark of consciousness that results from this kind of information integration. Therefore information integration and the coalitions of modules are considered by some researchers as necessary prerequisites for consciousness. According to this view consciousness arises when all modules coalesce and focus on the same entity; one unity, one consciousness [Tononi *et al.* 1998, Shanahan 2010].

The HCA fulfills the requirements of seamless multisensory and sensorimotor information integration and percept externalization by the associative interconnections between the modules (broadcast). However, the HCA allows also partial and concurrent coalitions of modules.

Why should concurrent coalitions be allowed or needed? Would these destroy the global unified focus and consequently prevent any possible consciousness? On the contrary. Consider this example: You, the respected reader of this book are now reading this page, this line and right now this word. Now you are reading the words that follow. Your eyes scan the text line by line, word by word and every now and then your hand turns the page. It is your brain that controls these operations, yet you are not really aware of these and why should you? After all, these operations do not contribute anything to the content of the text that you are reading. Nevertheless, the brain must provide resources also to these operations, sensory resources to see the text, motor resources to control the eyes and your hand when you turn the page. When these actions are considered in terms of module-based cognitive architectures, it will be noted that all these actions call for coalitions of several modules. In addition, these coalitions take place concurrently with those coalitions that allow the understanding the text. Thus the whole operation of book reading calls for concurrent coalitions of modules. In a similar way, most tasks that we do, call for concurrent coalitions of modules and the activity of many of these coalitions remain sub-conscious. It should also be understood that sensory modalities, like vision, may consist of a number of sub-modules, which may individually take part in different coalitions.

The HCA is a distributed architecture that due to its extensive cross-connections easily allows the formation of ad hoc coalitions, which may be called for by each situation. Only the action of those coalitions that are able to produce report and memory percepts will become consciously perceived. This fulfills one prerequisite for consciousness; the first and mandatory prerequisite for consciousness, the inner phenomenal appearance, is addressed by the direct and transparent perception process and the introspection via feedback loops as discussed before.

Summary

- The Haikonen Cognitive Architecture (HCA) is a dynamic perceptive system for the production of possibly conscious human-like cognition including the flow of inner imagery and natural language inner speech.
- The HCA utilizes associative information processing with distributed signal representations with emotional significance evaluation and match/mismatch/novelty detection as attention and relevance guiding factors. Information processing is executed by associative neural networks that allow the transition from sub-symbolic to symbolic processing.
- The HCA consists of perception/response feedback loop modules that broadcast their percepts to each other.

Chapter 19

Mind Reading Applications

19.1. Mind Reading Possible?

Would it be possible to develop technologies that would allow the reading and viewing other people's thoughts and inner imagery and perhaps recording these as a video? This kind of technology could have important applications in the diagnosis and care of brain injuries and especially in the care of *locked-in syndrome* patients. A locked-in syndrome patient is awake and aware, but is not able to move or communicate verbally due to the paralysis of most of the voluntary muscles. In addition to patient care, mental content imaging technology would also open a new window for brain research.

We experience our mental content in the form of qualia. It was argued earlier that qualia are not directly accessible to other people via any brain imaging technologies, as these can only detect and measure physical processes that are related to the brain activity and produce *symbolic representations* of these. Qualia are direct ways of information representation and symbolic representations are not qualia. Brain imaging methods do not detect qualia or mental content, instead they detect physical processes that may or may not be related to conscious mental content. Thus, the direct detection and recording of mental content would seem to be impossible.

However, it was also argued before that some of the brain activity constitutes the conscious mental content and the qualia are the way, in which the conscious brain activity appears to the subject. This, if true, would allow an indirect method for the imaging of mental content, because technologies for the detection and measurement of brain activity exist. Vice versa, the successful imaging of mental content would prove the equivalence of the physical brain processes and the mental content (and would also justify the main tenets of this book). This hypothesis has the following consequence: The data, the symbolic representations of the physical processes produced by various brain imaging technologies, would be related to the mental content and this content could be determined from this data. The crucial question of technological mind reading is: What kind of method could extract mental content from the data produced by the various brain imaging technologies? Surprisingly, the principle of this kind of a method is a simple one.

19.2. The principle of the Imaging of Inner Imagery

According to the perception/response feedback loop hypothesis, visual perception and internal imagination produce neural activities that occupy the same neural area. In the brain this area is the visual cortex that is located at the back of the brain (occipital lobe). The neural activity within the visual cortex would be related to consciously seen and imagined imagery and the detection of this activity would be the basis for the external imaging of inner imagery. Obviously, external non-invasive and painless methods would be preferred for the detection of the neural activity; additional peep holes in the skull are not a popular approach.

Localized neural activity in the brain can be externally mapped by the so-called functional magnetic resonance imaging (fMRI) technology. A fMRI scanner consists of a large magnet, a radio frequency (RF) pulse generator and a receiver. During fMRI scanning the test person is placed inside the cylindrical machine. Strong magnetic field is used to align the spins of hydrogen nuclei (protons) in the brain. A short RF pulse is used to change the spins and at that moment the protons absorb energy. When the pulse is turned off, the spins realign to the magnetic field and the protons emit the absorbed energy as a radio signal. The decay time of this radio signal depends of the local ratio of de-oxygenated/oxygenated blood, which is supposed to be related to the local neural activity. This is the so-called *BOLD effect* (Blood Oxygenation Level Dependent effect) that depends on the inhomogeneities in the magnetic field caused by oxygen level changes in the blood.

fMRI technology allows the detection of neural activity with the spatial resolution of few millimeters or even less and the temporal resolution of few seconds. This allows the computation of scan images of the brain and its activity, but these images as such do not represent any mental content or inner imagery of the scanned test person.

The next step in the imaging of the mental content with fMRI technology would be the interpretation of the scan images. More or less repeatable scan image patterns appear, when the test person views repeatedly the same or similar images. Statistical and associative methods can be used to find correlations between the seen objects and the scan image patterns. Thereafter probable images of the imagined objects can be constructed by these correlations. The principle of the associative reconstruction of mental content is depicted in Fig. 19.1.

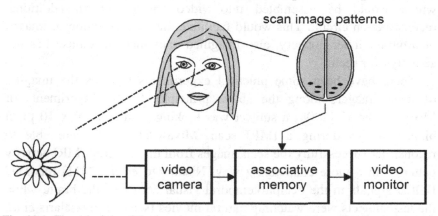

Fig. 19.1. The principle of the imaging of inner imagery. The scene that the test person sees is captured by a video camera and is stored in an associative memory. The test person's brain scan patterns are associated with the stored imagery and later on, when the test person imagines the same scenes, similar brain scan patterns are detected and these evoke associatively corresponding images at the associative memory. These images can be seen on a video monitor.

In Fig. 19.1 the subject and a video camera view the same object. The test person's visual cortex is scanned by the fMRI method and sequences of scan image patterns are achieved. The captured video images are associated with the brain scan patterns and are stored in the associative memory. The video images may be later on evoked associatively by similar brain scan patterns, when the test person is seeing or imagining similar objects. The evoked imagery can be seen on the video monitor.

The principle of the imaging of inner imagery is simple, but in practice there are some important points that should be noted. The eye has a small sharp vision area, the fovea, in the retina and this determines the instantaneous visual attention focus. The video camera should have a similar visual attention focus area and the direction of the camera should track the gaze direction of the test person's eyes at all times. This would guarantee that the foveas of the eyes and the visual attention focus area of the camera would capture the same objects at each moment.

Furthermore, visual pre-processing should be used. Preferably, this preprocessing should extract similar visual features as the brain does. Consequently, the associative memory should contain visual features, which would be assembled into video images in an additional reconstruction circuit. This would facilitate the reconstruction of images of arbitrary inner imagery, also imagined ones that the subject has not actually ever seen.

There have been some practical experiments towards the imaging of inner imagery along the above principle. In the experiments of Miyawaki *et al.* [2008] a subject was looking at simple 10 x 10 pixel binary images during a fMRI scan. Miyawaki *et al.* were able to reconstruct successfully the seen images from the detected fMRI activity patterns. In a more challenging way, Nishimoto *et al.* [2011] recorded BOLD signals in the occipitotemporal visual cortex of the brain, while the test subjects were watching natural movies (see also [Naselaris *et al.* 2009]). It was found out that with some processing, the BOLD-signal patterns correlated well with the information in the movies. A Bayesian decoder was also constructed and used to translate the BOLD signal patterns into video images. According to Nishimoto *et al.*, remarkable reconstructions of the viewed movies were produced and the results would seem to verify that current fMRI technology is able to allow the decoding of dynamic brain activity.

The works of Miyawaki *et al.* and Nishimoto *et al.* would seem to confirm the hypothesized connection of the physical brain processes and the mental content and also the hypothesis that the imagined and sensorily perceived images occupy the same neural area. These hypotheses are also basic tenets of this book and the basic idea behind the perception/response feedback loop model.

19.3. The Perception/Response Feedback Loop in the Imaging of Inner Imagery

The imaging of inner imagery is based on the correspondence between the visual sensory stimuli and the detected brain activity patterns that are caused by the perception of these sensory stimuli. This correspondence can be found out by subjecting the test person to well-defined visual stimuli and recording these and the related brain activity patterns at the same time. In this way two tracks will be created, the brain activity track and the matching visual stimuli track. It is then assumed that the test person's imaginations will evoke similar brain activity patterns as the actual sensory perception of the same. In other words, the imaginations would correspond to the sensory stimuli. The sensory stimuli that corresponds to certain brain activity patterns could be determined from the created brain activity track and the matching stimuli track. This sensory stimuli should then indicate and represent topic of the test subject's imaginations.

The practical realization of the above principle is not without problems. First, the sensory stimuli in the form of images and sounds should be matched with the corresponding brain activity patterns by recording these simultaneously. However, it may happen that the recorded stimulus images and sounds are not exactly those that have been consciously attended to by the test person. If they are not, the recorded stimuli track and the brain activity track will not correspond to each other. Therefore some mechanisms that would synchronize the focus of the stimulus image to the attention focus of the test person would be needed.

In practice, sensory stimuli are almost always somewhat different even when the same objects and events are perceived. Also the brain activity patterns differ. To overcome this problem, some kind of generalization and classification process would be required.

The use of the perception/response feedback loop principle will help to solve some of the above problems. Information processing with distributed feature signal representations and associative neurons would provide means to match the detected brain activity patterns and to evoke the corresponding imagery in the form of video images. The SOFT-AND function of the associative neurons may provide useful generalization and classification, if the features to be depicted by the signal representations are properly chosen. An ideal perception/response feedback loop for the imaging of inner imagery would be the one that closely matches the visual preprocessing of the brain.

The use of the perception/response feedback loop in the imaging of inner imagery is depicted in Fig. 19.2.

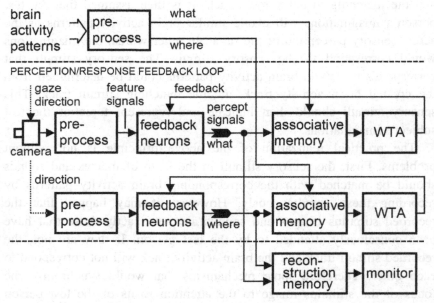

Fig. 19.2. The perception/response feedback loop in the imaging of inner imagery. Brain activity patterns are associated with corresponding visual percepts. Detected brain activity feature signals evoke corresponding visual "what" and "where" percept signals and these are transformed into viewable images in the reconstruction memory.

The system of Fig. 19.2 has two modes of operation, which are the learning mode and the imaging mode.

In the learning mode, the perception/response feedback loop associates detected brain activity patterns with corresponding visual information from the video camera. The image captured by the video camera is preprocessed into distributed feature signal arrays that depict elementary visual features at each position in the focus area of the visual scene. In this application the video camera should have a narrow field of acute vision that would match the fovea of the human eye. The gaze direction of the eyes of the test person would be detected and the camera direction would be constantly adjusted to track the detected gaze direction. In this way the visual focus of the eyes and the camera would always coincide.

At the same time, the brain activity patterns (e.g. from fMRI) are preprocessed into distributed signal representations of "what" and "where". The visual features in the visual attention focus area would give the "what" information and the gaze direction would give the "where" information. These representations are forwarded to the associative memories in the perception/response feedback loop. In the learning mode, these representations are associated with the corresponding representations from the video camera.

In the imaging mode, the detected brain activity patterns evoke corresponding visual "what" and "where" information at the associative memories. This information is transformed into viewable images in the reconstruction memory. In this mode, the camera and the related preprocesses are disabled and do not contribute to the overall operation.

A prerequisite for the correct operation of this system is the proper design and operation of the reconstruction memory. This can be checked by the direct reconstruction of the camera image. The reconstruction memory must be able to transform the distributed representations from the camera, the visual percepts, into viewable video. If the direct reconstruction does not work, then the reconstruction of mental imagery will also fail. A personal computer may be used as the reconstruction memory and the display.

19.4. Inner Speech and Unvoiced Speech

The object of mind reading is not only the imaging of mental imagery. Another obvious target is the listening of the thought flow that appears in the form of the inner speech. The general principle of the imaging of the inner imagery can be applied to this task, too. The brain activity signals that could be used in this application would be those that control the production of speech. These signals should be found at the left side of the brain, at the so-called Broca area. However, the use of fMRI for this purpose would be challenging due to the limited temporal resolution of this method.

It has been found out that inner speech generates speech motor command signals even when the thoughts are not voiced aloud [MacKay 1992]. This phenomenon offers another possibility for the detection of inner speech. The speech motor commands go to the laryngeal area (neck) muscles that are responsible for the production of speech sounds. Activated muscles produce low-level electrical potentials that can be externally detected on the surface of the skin. This technique is known as *surface electromyography* (EMG).

Useful speech-related EMG signal patterns can be detected by the use of several properly placed electrodes at the neck and chin area. Strong signals are generated during voiced speech, but also almost spoken unvoiced speech and possibly the inner speech as such are able to generate detectable signals. It has been demonstrated that both voiced and unvoiced words can be determined from these signal patterns [Wand & Schultz 2009].

19.5. Silent Speech Detection with the Perception/Response Loop

The perception/response feedback loop can be utilized also for the listening of unvoiced speech with EMG signals. Inner speech may also be recognized by the same system, if the sensitivity of the EMG signal detection is sufficient and the preprocess is able to cope with the lower signal-to-noise ratio caused by the weaker EMG signals. The principle of a perception/response feedback loop based system for the listening of silent speech is depicted in Fig. 19.3.

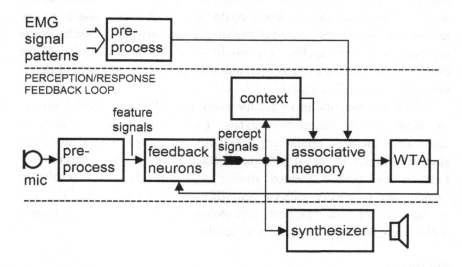

Fig. 19.3. The perception/response feedback loop system for the listening of silent speech.

The system in Fig. 19.3 has two operation modes; the learning mode and the silent speech listening mode. In the learning mode, phoneme-related feature signals are extracted from the microphone signal and these are associated with the preprocessed EMG signal patterns. Speech has temporal dimension, therefore the associative memory must be a serial one, accepting strings of signal patterns. Here autoassociation is used in addition to heteroassociation. This will allow the utilization of the context correlation between adjacent phonemes.

  In the listening mode, the detected EMG signal patterns evoke strings of phoneme-related related feature signals. The most strongly evoked signals are selected by the Winner-Takes-All (WTA) threshold at the output of the associative memory. These signals are returned to the feedback neurons, that produce percept signals. The percept signals are forwarded to the context memory and the speech synthesizer. The context memory contains a constantly updated string of the previous phoneme percept signals, which are used as a partial cue for the next phoneme. This will utilize the correlation between adjacent phonemes and will in this way improve the recognition process. In a rather similar way even more advanced context could be used, if a complete cognitive architecture were used instead of the simple auditory perception/response feedback loop. This architecture would allow real understanding of the content matter of the inner speech and this, in turn, would help to detect correctly even weaker signals.

Some experimental proof for the viability of the associative approach in silent speech recognition exists. Lesser has used the author's associative memory [Haikonen 2007] for the recognition of unvoiced speech with EMG signal patterns [Lesser *et al.* 2008]. According to Lesser, this method yielded similar and partially better results in comparison to earlier pattern classification based methods. The full application of the perception/response feedback loop cognitive architecture should give even better results.

Summary

- Localized neural activity in the brain can be externally mapped by the so-called functional magnetic resonance imaging (fMRI) technology.
- Visual cortex fMRI activity patterns correlate with seen imagery and consequently, seen and imagined imagery can be deduced from the fMRI activity patterns.
- Unvoiced speech can be detected in a similar way from surface electromyography (EMG) signal patterns.
- The perception/response feedback loop can be used in these applications.

Chapter 20

The Comparison of Some Cognitive Architectures

20.1. Introduction

As illustrative examples of typical issues that relate to cognitive architectures, the following cognitive architectures are compared: the Baars' Global Workspace architecture, the Shanahan Global Workspace architecture and the Haikonen Cognitive Architecture. Of these, the Baars model is the oldest, the Haikonen model is next and the Shanahan model is the most recent. The Baars model is well-known and many other models have been derived from its general principles, such as the Baars-Franklin computational LIDA model [Baars and Franklin 2009], the CERA-CRANIUM model [Arrabales, Ledezma and Sanchis 2009] and recently, the Shanahan [2010] model.

Also the Haikonen model has influenced some developments in this area, for instance the model of Kinouchi [2009]. The Kinouchi model is remarkable, because it is one of the few that explicitly consider the physical layer i.e. the executing artificial neural machinery and the phenomenal layer, "the logical layer".

The architectures of Baars, Haikonen and Shanahan share some common ideas and features, but there are also some decisive differences. All these architectures are claimed to offer some kind of explanation for the phenomenon of consciousness or at least for the difference between conscious and non-conscious operation.

In the following, the main principles and features of these cognitive architectures are summarized and compared with each other. Their ability to explain the real problem of consciousness, that is, the mechanism for the generation of the internal subjective appearance of neural activity in the form of qualia, is also compared. This is the most important and decisive point; a cognitive architecture that does not address this issue properly, is not a sufficient model for the mechanisms of a conscious mind.

20.2. Baars Global Workspace Architecture

20.2.1. *Baars Concept of Consciousness*

Baars sees the human consciousness as a biological adaptation that allows the brain to learn, interpret and interact with the world [Baars 1997]. Baars believes that the workings of the brain are distributed. There is no central command, instead the various networks of the brain are controlled by their own aims and contexts. All this has to be organized somehow and for that purpose there would be a network of neural assemblies that would display the contents of consciousness. Baars continues to propose that the cortical sensory projections areas would be best candidates for these networks. According to Baars the contents of consciousness include the perceptual world, inner speech and imagery, traces of immediate memory, feelings, autobiographical memories, intentions, expectations, and so on. Baars proposes the concept of *focal consciousness*; having a mental event *that can be reported*, such as seeing an object or having a thought.

Baars proposes that consciousness has a function; it creates access to the information in various locations in the brain. Baars sees *consciousness as the publicity organ* of the brain. This organ would facilitate accessing, disseminating and exchanging of information. This organ would also exercise global coordination and control. Baars goes further: Consciousness prioritizes percepts; is related to problem-solving by providing access to unconscious resources; facilitates decision making; optimizes the trade-off between organization and flexibility; helps to recruit and control actions; is needed for error detection and editing of action plans [Baars 1997, pp. 157–164].

20.2.2. *Baars Model*

The Baars Global Workspace Model [Baars 1988, 1997] is based on the idea of a network of neural assemblies that would display the contents of consciousness. This network would be a central working memory area, the so-called *Global Workspace*. The proposed workspace would act as a theater stage that would display and broadcast the contents of consciousness to the unconscious audience. This unconscious audience would consist of operators like memory systems, interpretation and recognition, action control, skills, etc. The Baars Global Workspace Model is depicted in Fig. 20.1.

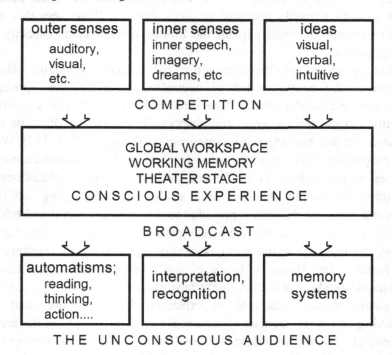

Fig. 20.1. Baars Global Workspace Model (compiled after [Baars 1997]). The global workspace working memory is a kind of a theater stage that receives information from outer and inner senses and results from mental activity. The "spotlight of attention" selects the contents of the theater stage that is broadcast to the "unconscious audience". It is proposed that the broadcast part of the contents of the global workspace is also the contents of consciousness.

In the Baars model the contents of consciousness originate from outer senses (eyes, ears, touch, etc), inner senses (introspection of inner speech, imagery etc.) and ideas. Transmissions from these sources must compete against each other for the access to the global workspace. By definition, the contents of the global workspace memory constitute the conscious experience. The Baars model of conscious cognition is a *theater model*.

In the Baars model the global workspace working memory is a kind of a theater stage. Only a part of contents of the theater stage is consciously perceived at any time, namely the part that is "illuminated" by the "spotlight of attention". The rest of the contents of the theater stage would be readily available to consciousness. Baars sees that this kind of operation would explain the limited instantaneous capacity of consciousness.

Baars proposes also that there are so-called context operators behind the scene; the action would be controlled by a director, spotlight controller and local contexts. The director, which performs executive functions, is the self, an agent and observer [Baars 1997 p. 45]. The self is the entity that has access to consciousness [Baars 1997 p. 153]. With this proposition Baars comes perilously close to the discredited concept of *homunculus*, "a little human" inside the brain. In early explanations of the mind the homunculus was seen as the necessary acting self that observed whatever the senses provided to it to be seen, heard and felt. It should be obvious that this is not a logically satisfactory explanation at all, as the mystery of the mind is merely replaced by the mystery of the homunculus. Baars is aware of this problem and argues that a homunculus-style explanation might nevertheless be useful, if it reduces the extent of the matters to be explained. Baars proposes that the observing self would consist of pattern recognition circuits in the brain. In general, the self would be a mental framework that would remain stable across the course of life [Baars 1997, pp. 142–153]. The self would be able to will and plan, prioritize, make decisions and solve problems. According to Baars, it is consciousness that facilitates these functions. It is not immediately obvious that this reasoning is free of any *circularity*, the act of explaining things by themselves.

The Global Workspace Model is also a *Blackboard Model*. Baars [Shanahan and Baars 2005] states that the global workspace architecture was inspired by and is derived from earlier Artificial Intelligence blackboard systems like those presented by Hayes-Roth [1985] and Nii [1986]. In Artificial Intelligence the Blackboard Model is a method of computation with a group of specialist operators. A problem is "written" on the "blackboard", a common working memory, where it is available for all the specialist operators. Each operator reads the blackboard and tries to update the blackboard presentation whenever it can provide a step towards the final solution. However, at each moment only the specialist that has the most relevant update, may write on the blackboard; therefore the specialists must compete against each other for the access to the blackboard.

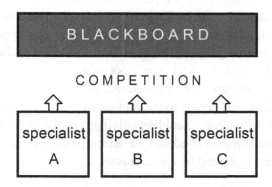

Fig. 20.2. The Blackboard system. The blackboard displays the problem at hand and also the intermediate steps towards a solution. The specialist that has the best contribution towards the solution at each moment may write its contribution on the blackboard for everybody to see.

The blackboard model is depicted in Fig. 20.2. At each moment any of the specialist operators A, B, C may write on the blackboard if its contribution is more relevant than the contributions of the other specialist operators. (The number of specialist operators is not limited to the three in this figure in actual applications.) The blackboard "displays" or "broadcasts" continuously its contents to every specialist operator. It can be seen that the blackboard method has an important restriction: On the blackboard the information must be presented in a common language or code (lingua franca) that every specialist operator can understand. However, in their internal processes the specialist operators may use whatever codes are suitable.

20.2.3. *What Is Explained by the Baars Model*

According to Baars, *consciousness* is created by the broadcasting of the information selected by the "spotlight of attention" that "illuminates" some of the contents of the theater stage, also known as the global workspace working memory.

The difference between conscious and non-conscious activities is explained as follows: Most of the mental activity and contents of the brain is non-conscious. Only the part of the contents of the theater stage, which is illuminated by the spotlight of attention is conscious.

The Baars model proposes an explanation to the apparent *limited capacity, vast access* property of consciousness. The global workspace has a limited capacity and the spotlight of attention limits the instantaneous contents of consciousness even more. However, consciousness has vast access to the non-conscious contents of the brain.

Percept prioritization, problem-solving, decision making, action control, error detection and editing of action plans, etc. are explained as functions that are executed by consciousness.

20.2.4. *What Is Not Explained by the Baars Model*

The Baars model is a general one, which does not give details of the actual workings of the underlying machinery. In the brain the machinery would consist of neurons and neuron groups, but the Baars model does not define these, nor propose any detailed method of interaction between these.

The Baars' Global Workspace model does not explain, why and how the neural activity could appear internally in the form of qualia. Thus the Baars Model does not address the fundamental issue of consciousness at all and therefore is not a sufficient model for conscious agents.

Summary of Baars Global Workspace Architecture

- The Baars Global Workspace Architecture is a theater model.
- The Baars Global Workspace Architecture is derived from earlier Artificial Intelligence Blackboard models.
- The Baars Global Workspace Architecture consists of a number of specialist modules and a common working memory area, the so called Global Workspace.
- The Global Workspace broadcasts its contents to all specialist modules.
- Each specialist module competes for the right to transmit its contribution to the Global Workspace; only the most relevant contribution is accepted at a time.
- The Global Workspace model explains the difference between conscious/non-conscious activity.
- The Global Workspace model explains the serial narrow bandwidth nature of the stream of consciousness; specialist modules operate subconsciously in parallel way, while the stream of consciousness mediated by the global workspace is serial.
- Baars' consciousness has a function; it facilitates accessing, disseminating and exchanging information and exercises global coordination and control.

20.3. Shanahan Global Workspace Architecture

20.3.1. *Shanahan Concept of Consciousness*

Rather than defining consciousness, Murray Shanahan [2010 p. 67] tries to approach the phenomenon by inspecting the difference between conscious and unconscious processes. Shanahan sees that unconscious actions are kinds of automatic actions that are executed without conscious attention. On the other hand, conscious actions are characterized by introspective reportability, enhanced flexibility in novel situations, the mental ability to execute problem-solving steps and the ability to lay down memories.

Shanahan sees that in the brain there are many simultaneously ongoing activities. According to Shanahan, an important hallmark of the conscious condition is *unity*, the integration of these different brain activities so that they can influence each other. Shanahan proposes that in the conscious condition, the brain's multitude of processes act as an integrated whole. Shanahan sees that in the brain there is a mechanism that facilitates this integration, which leads to the conscious condition; this mechanism is the global workspace. Shanahan's global workspace is, however, different from that proposed by Baars.

20.3.2. *Shanahan Model*

Shanahan [2010] has presented an alternative Global Workspace model, which is inspired by the Baars model. The basic components of the Shanahan and Baars models are the same; a number of specialist modules and a global workspace. The specialist modules would try to broadcast their information to the global workspace. However, *Shanahan recognizes that the theater models of mind border on the concept of the discredited homunculus.* Therefore he rejects the theater models of consciousness and, accordingly, the role of the global workspace as a theater stage. Consequently, the Shanahan architecture is not a theater model.

In the Baars model the global workspace is a working memory that acts as a theater stage, where the conscious contents of the mind is displayed to the audience of the non-conscious modules. Shanahan rejects this and proposes that instead of a theater stage and a common working memory area, the global workspace should be thought as a communications infrastructure. This infrastructure would connect the various autonomous specialist units to each other and in this way would facilitate the brain activity integration that is a supposed prerequisite for consciousness [Shanahan 2010 p. 111].

However, in the Shanahan model the global workspace is more than a mere communications infrastructure as it executes also other functions. Shanahan proposes [Shanahan 2010 p. 148] that besides being the locus of broadcast, the global workspace is also the arena for competition between rival connections of modules. The essential elements of the Shanahan model are depicted in Fig. 20.3.

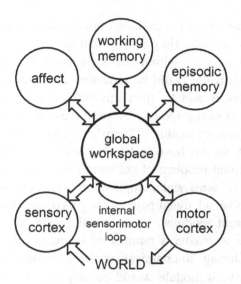

Fig. 20.3. The Shanahan model (compiled after [Shanahan 2010]). The Shanahan global workspace is a communications infrastructure that connects the various modules with each other. The modules must compete for the access to the global workspace. An internal sensorimotor loop allows the mental simulation of actions. It is proposed that the information that goes through the global workspace constitutes the contents of consciousness.

Figure 20.3 depicts the main modules and interconnections of the Shanahan model. In this model the global workspace infrastructure connects the modules of affect, working memory, episodic memory, sensory cortex and motor cortex to each other. Shanahan assumes that this division of functions would roughly conform to the anatomical divisions in the brain. The sensory cortex and the motor cortex form a functional loop, which is closed by the external world; the actions executed by the motor cortex can be perceived by the senses and the sensory cortex.

Hesslow [2002] has proposed that thinking is simulated action and perception, which is enabled by an internal sensorimotor feedback loop.

Shanahan has augmented his model by the inclusion of an internal sensorimotor loop in the style of Hesslow. This loop would allow the internal perception of planned actions without the need to actually execute these.

Shanahan is aware of the common code problem that is inherent in global workspace models. He proposes that the need for a *lingua franca* [Shanahan 2010 p. 118] within the framework of global workspace communication structure could be remedied easily. If a module A is to influence modules B and C, then this influence can be mediated by different signals; i.e. the module A would use dedicated codes when transmitting to different modules. Shanahan states: "The signals going to B and C from A do not have to be the same". Obviously this method bypasses the original problem of common code, but at the same time it creates further problems and complications. The modules cannot now broadcast one universal signal pattern, they have to send different signal patterns to different modules. Consequently, each transmitting module would now have to master a number of languages instead of a single universal one. During communication the transmitting module would have to know, which module would be targeted and it would have to generate a specific signal pattern for that module. However, associative communication could also be utilized in Shanahan's model. This would remedy the common code problem.

20.3.3. *What Is Explained by the Shanahan Model*

Shanahan proposes that the global workspace communication infrastructure explains the conscious/unconscious distinction in a way that is compatible with the ideas of Dennett and Tononi. According to Dennett [2001] the difference between the conscious and unconscious processing would arise from the difference between local and fully global influence; a conscious process would recruit global resources. Tononi's [Tononi *et al.* 1998] information integration theory states that a conscious process would involve global interactive neural activity.

Along these lines, Shanahan proposes that the information mediated by the global workspace communication infrastructure constitutes the contents of consciousness, while the specialist units operate unconsciously. The global workspace communication infrastructure would allow an integrated response and it would enable learning and episodic memory making, etc. by allowing the various modules to cooperate on the same topic. Shanahan states: "Perfect integration occurs when the being as a whole is brought to bear on the ongoing situation" [Shanahan 2010 p. 112]. The limited bandwidth of the global workspace communications infrastructure would allow only one coalition of processes at a time. This limitation would direct the focus of attention onto a single, unified object and lead to the forming of a single, unified thought. In this way the stream of serial consciousness would arise from the parallel processes of the brain.

Unfortunately, the allowing of only one coalition at a time comes at a cost. In addition to the ongoing conscious experience, conscious humans are able to process additional semiconscious or sub-conscious tasks at the same time. There may be some sub-consciously processed tasks within the various modules, but there are also tasks that call for limited coalitions of modules. For instance, the reading of a text calls for sensorimotor coalitions that control the visual attention and the movement of the eyes. However, this action is not consciously perceived, unless the reader specially focuses attention on this. Yet, at the same time, other coalitions must process the meaning of the text; the results from these coalitions would be consciously perceived. In Shanahan's model this could not happen; simultaneous, but separate coalitions are forbidden.

The internal sensorimotor loop explains the ability to imagine motor actions as if they were actually executed. This function would allow the planning and selection between suitable actions.

20.3.4. *What Is Not Explained by the Shanahan Model*

The Shanahan Model does not explain, why and how the supposedly conscious neural activity in the global workspace communications structure could appear internally in the form of qualia. Thus, the Shanahan Model does not address the fundamental issue of consciousness and therefore is not a sufficient model for conscious agents.

Summary of The Shanahan Global Workspace Architecture

- The Shanahan Global Workspace Architecture is not a theater model.
- The Shanahan Global Workspace Architecture consist of a number of specialist modules that are connected to each other via a communications infrastructure, the Shanahan global workspace.
- The communications infrastructure is a narrow bandwidth connective core.
- The specialist modules must compete for the access to the connective core.
- Specialist modules operate non-consciously. Consciousness occurs when the communications infrastructure mediates an integrated response by bringing the whole system to bear on the ongoing situation.
- The information mediated by the global workspace communication infrastructure constitutes the contents of consciousness.
- The Shanahan model includes an internal sensorimotor loop that allows the imagination of motor responses.

20.4. Haikonen Cognitive Architecture

20.4.1. *Haikonen Concept of Consciousness*

The Haikonen cognitive architecture is based on the following hypotheses about consciousness:

- Consciousness is the presence of qualia-based *internal appearances* of the direct and externalized sensory percepts of the environment and the body and the virtual percepts of mental content.
- Reportable percepts constitute the contents of consciousness; without percepts there is no consciousness.

- A conscious mind has the flow of sensory and imagined percepts and possibly a meaningful flow of symbolic percepts such as a natural language inner speech.
- A conscious mind has the ability to report its contents to itself and outside in various ways and possibly by using a natural language.
- Consciousness is an inner appearance of information, it is not an agent.

Higher symbolic cognition and the mastery of a natural language are not seen as necessary prerequisites for consciousness. However, it is recognized that a plausible model of cognition and consciousness should be able to explain also these.

20.4.2. *Haikonen Model*

The Haikonen Cognitive Architecture is primarily a blueprint for artificial robot brains. As such, it details both the overall architecture, the information processing method and the basic building blocks. It is inspired by the human brain and cognition, but does not seek to model the brain accurately.

The Haikonen Cognitive Architecture tries to address the qualia-aspect of consciousness; it tries to create inner appearances of the perceived world via direct sensory perception and also inner appearances of selected mental content via the feedback principle. Technical details of the Haikonen model are given in [Haikonen 2007]. More general details as well as cognitive and philosophical background can be found also in [Haikonen 2003].

The Haikonen Cognitive Architecture is designed to fulfill the requirements of perception and sensorimotor integration, the flow of sensory and imagined percepts, emotional effects and a natural language. In robotic implementations it has means to perceive its environment and body, introspect and report its mental content, understand situations and their requirements, plan and control motor actions and exercise judgment with emotional values.

The Haikonen architecture is a parallel and distributed one, without any global workspace or "a theater stage". The executive and attention functions are distributed. The architecture consists of a number of modules for sensory and motor modalities. These modules are rather similar to each other and are based on the principle of the perception/response feedback loop. Each sensory modality has its own perception/response feedback loops that can handle a large number of parallel signals.

20.4.3. *What Is Explained by the Haikonen Model*

Qualia: The Haikonen model is founded on the proposition that consciousness is based on perception. Direct perception, as opposed to indirect perception via symbols, is assumed to facilitate some kind of qualia. The perceptual information is carried by neural signals, but the system is not able to perceive the signals as such; consequently only the perceptual information is causing effects and these constitute the machine qualia. These qualia are not necessarily similar to the human qualia, with the possible exception of amodal qualia.

The difference between conscious and non-conscious activities is explained as follows: The associative cognitive system has continuously ongoing activities; some of these are related to sensory perception and appear as percepts. Some activities are related to forking chains of associations "deep" within the system. Reportable percepts are conscious while the association chains within the system are not. The products of the non-conscious activities may become consciously perceived when they are transformed via the feedback loops into the equivalents of sensory percepts.

The reportability aspect of consciousness is explained as follows: Percepts are broadcast to every other modality; the receiving modality determines which broadcasts are accepted and attended to. When percepts of a certain event are accepted by many modalities, multiple associative connections may be formed and consequently memories of that event will be created. This allows the reporting of that event in terms of the receiving modalities. Thus attended percepts will be remembered for a while and can be reported; this is one criterion for conscious percepts.

The serial nature of consciousness is explained as follows: Basically, the associative system is a parallel network that is able to process many things simultaneously. However, in the conscious mode the various modules are focusing their attention on the same thing. Only one thing (or few things depending on the actual realization of the system) can be globally attended to at a time, therefore the flow of consciously attended percepts will be temporally serial.

Cognitive functions: The Haikonen model proposes means for the artificial realization of a number of cognitive functions, such as perception, learning, memory making, introspection, language and inner speech, imagination, emotions.

Summary of the Haikonen Cognitive Architecture

- The HCA is not a theater model or a global workspace model.
- The HCA seeks to incorporate phenomenal perception, qualia.
- The HCA is a sub-symbolic/symbolic neural system.
- The HCA consists of a number of specialist modules; perception/response feedback loop units, that are connected to each other via wide bandwidth signal lines.
- The specialist modules broadcast their percepts.
- The specialist modules decide when to accept broadcasts and from whom.
- Specialist modules operate non-consciously.
- When modules focus on the same object and situation ("information integration"), allowing cross-association and evocation, the reportability aspect of consciousness is realized.
- The HCA incorporates inner speech.
- The HCA incorporates system reactions that facilitate pleasure and pain according to the SRTE model.

20.5. Baars, Shanahan and Haikonen Architectures Compared

The Baars, Shanahan and Haikonen architectures utilize a number of non-conscious autonomous specialist modules that interact with each other in one way or another. In the Baars model this interaction takes place via the global workspace memory, the "theater stage", which broadcasts the attended contents to the specialist modules. In the Shanahan model the interaction takes place via the global workspace communications structure and in the Haikonen model the modules communicate directly with each other.

Figure 20.4 depicts the general principle of the internal communication between the autonomous modules and the global workspace working memory in the Baars model. The global workspace working memory broadcasts the attended part of its contents to each autonomous module. The modules must compete against each other in order to be able to post their information to the global workspace working memory and eventually via this route to the other modules. The number of the autonomous modules is not limited to the four modules A, B, C, and D shown in this depiction. Further modules would be connected to the global workspace in a similar way.

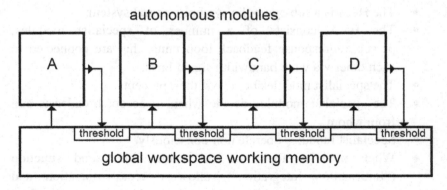

Fig. 20.4. Internal communication in the Baars Model. The transmissions from the autonomous modules A, B, C, D compete against each other for the access to the global workspace working memory. The global workspace broadcasts its contents to every autonomous module.

Figure 20.5 depicts the general principle of the internal communication between the autonomous modules and the global workspace communications structure in the Shanahan model.

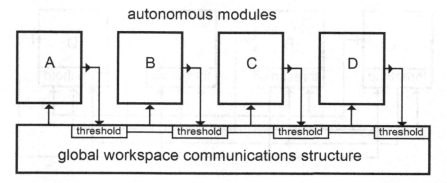

Fig. 20.5. Internal communication in the Shanahan Model. Transmissions from the autonomous modules compete against each other for the access to the global workspace communications structure. The global workspace broadcasts its contents to every autonomous module.

The modules of the Shanahan model must compete against each other for the right to input information the global workspace communications structure. This information is received by the other modules from the global workspace communications structure. The global workspace communications structure has limited bandwidth and allows the formation of one coalition of processes at a time, the system cannot support several simultaneous coalitions.

Figure 20.6 depicts the general principle of the internal communication in the Haikonen model. In the Haikonen model autonomous modules broadcast their information directly to each other. A receiving module determines, which (if any) of the simultaneously available transmissions is accepted at each moment. A module may also solicit a certain response by transmitting suitable cues that may evoke a strengthened response and transmission in the solicited module.

In the Haikonen model, simultaneous separate coalitions of modules are allowed and these ad hoc cooperative coalitions may form as the need arises. This property facilitates the execution of several tasks at the same time. However, only one task of these may be executed in a way that displays the hallmarks of consciousness.

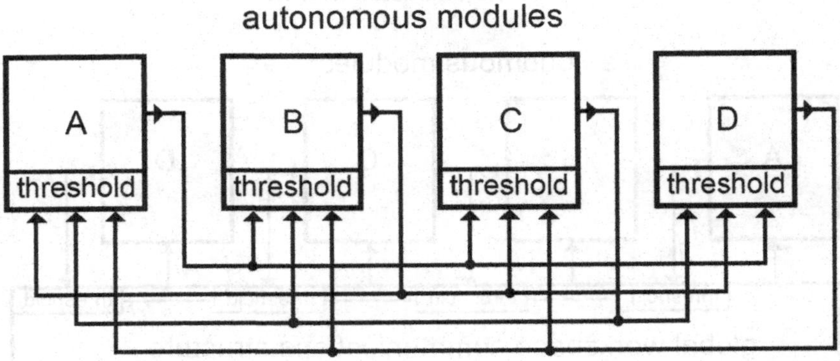

autonomous modules

Fig. 20.6. Internal communication in the Haikonen Model. Each autonomous module broadcasts its output to every other autonomous module. Each module has its own input threshold that determines which (if any) of the simultaneously available broadcasts is accepted at each moment. Arbitrary "ad hoc" coalitions of modules are possible.

In the Baars and Shanahan models the global workspace can be understood as the site of conscious mental content. The Haikonen model does not have a specific site for conscious mental content and the reportability hallmark of consciousness arises from the cooperation between modules. Consequently, in the Haikonen model there is no box that could be labeled as "consciousness"; every module operates in the same way whether the overall action is conscious or not.

The Baars, Shanahan and Haikonen models propose that the focus of consciousness is determined by global attention. However, the Haikonen model does not utilize a global workspace, which is seen as redundant, because the modules can communicate and compete directly with each other. There are also other important differences. Baars and Shanahan models do not address the real problem of consciousness; how the inner subjective experience arises from their processes. They do not really explain qualia, the inner appearance aspect of consciousness. They do not specify how inner speech could arise, either. The Haikonen model is specific on these issues.

Chapter 21

Example: An Experimental Robot with the HCA

21.1. Purpose and Design Principles

The Experimental Cognitive Robot XCR-1 is designed as a test bed for machine cognition experiments that involve sensorimotor integration with motor action generation and control using the Haikonen Cognitive Architecture (HCA) [Haikonen 2010, 2011]. The main design purpose of the XCR-1 was to create a simple robot that would utilize neural associative processing with the HCA. The robot should prove that the design principles would work also in minimalist realizations.

The robot XCR-1 is an autonomous, small three-wheel robot with gripper arms and hands, and simple visual, auditory, touch, shock and petting sensors. It recognizes some spoken words and has a limited vocabulary self-talk.

The sub-symbolic/symbolic processing properties of the HCA cannot be fully produced by computer programs. The HCA is also a parallel processing system and this is another benefit that would be lost with the use of microprocessors. Therefore the XCR-1 utilizes hardwired neural circuits instead of the more common program-driven microprocessors and possible links to a master computer, which would execute the more demanding computations. Thus, the XCR-1 is not program-driven and represents a completely different approach to self-controlled robots.

The XCR-1 has a set of motor action routines and hard-wired reactions to stimuli. These routines and the hard-wired reactions can be combined in various ways depending on the situation. Cognitive control may override reactions and modify the robot's behavior.

203

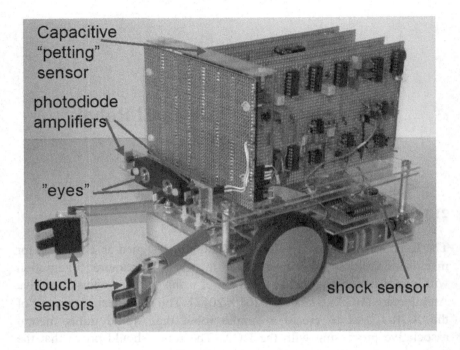

Fig. 21.1. The XCR-1 robot has three wheels for motion, gripper arms and hands with touch pressure sensors, and visual sensors, a microphone for sound perception, a shock sensor that detects mechanical shocks anywhere in the body and a petting sensor. The robot utilizes three small audio quality DC-motors.

21.2. Architecture

The robot XCR-1 utilizes a cognitive architecture that is a scaled-down version of the Haikonen Cognitive Architecture (HCA). The XCR-1 has the following functions: Target search, detection, verbal identification, approach and gripping; Functional effects of pain and pleasure; Association of emotional values with objects; Motivation by emotional values; Simple natural language (English) verbal report of the robot's internal states (self-talk); Limited speech recognition; Limited verbal learning.

The block diagram of the robot XCR-1 is given in Fig. 21.2.

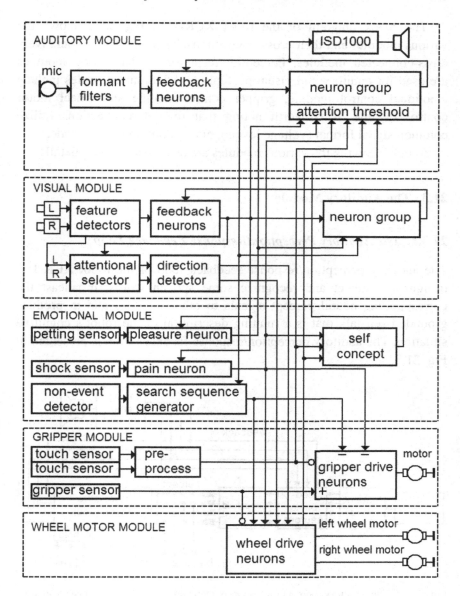

Fig. 21.2. The block diagram of the robot XCR-1 shows the auditory, visual, emotional, gripper and wheel motor modules and their cross-connections.

The block diagram of the robot XCR-1 in Fig. 21.2 shows the various modules and their cross-connections. There are five associatively cross-connected modules. Perception/response feedback loop modules are used for auditory and visual perception. Additional modules include emotional control module, gripper control module and wheel motor control module. It is worth noting that the interconnections utilize common signal format (voltage levels), but no common neural code.

In the following the various modules are described in more detail.

21.3. The Auditory Module

21.3.1. *The Auditory Perception/Response Feedback Loop*

The auditory perception/response feedback loop of the robot XCR-1 is designed to detect and recognize some spoken words, broadcast the corresponding neural percept signals to other modules and to produce grounded self-talk that is a running description of the robot's cognitive situation. The auditory perception/response feedback loop is presented in Fig. 21.3.

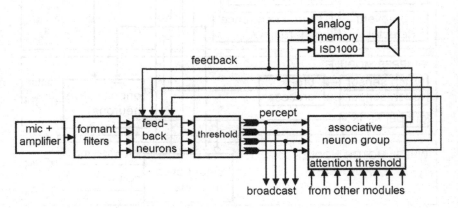

Fig. 21.3. The auditory perception/response feedback loop detects some words by formant filtering and transforms percepts from other modules into inner speech words, which are transformed into heard natural language words by the ISD1000 circuit.

A small electret microphone is used for the detection of sounds. The signal from the microphone is amplified and forwarded to formant filters that allow the detection of some spoken words. The detectable words generate patterns of neural signals, which are forwarded to the feedback neurons. The threshold circuit selects the actual percept signal pattern, which is broadcast to other modules.

The associative neuron group accepts broadcast percept signals from other modules via an attention threshold circuit. The accepted broadcast signals evoke associatively neural word signal patterns, which are fed back to the feedback neurons and are also forwarded to the speech producing analog memory circuit (ISD1000).

21.3.2. *Speech Recognition*

Spoken word detection is based on formant filters that detect the spectrum density of the sounds. Formant filters are band-pass filters that have their band-pass ranges tuned to suit the formant frequencies of the human speech. These formant filters include a rectifying and smoothing circuit that outputs a DC-voltage that indicates the signal intensity within the detected formant frequency band. Different combinations of formant frequencies correspond to different vowels. This process allows the recognition of the words regardless of their pitch.

In this simple application only vowel detection is used. This process is very limited, but suffices for the detection of few words and for the demonstration of the basic operational principles. It would be a straight forward task to extract additional auditory features by the addition of further formant filters, transient detectors and sequence detection, but in this simple case this would complicate the circuitry in an unnecessary way.

21.3.3. *Speech Production*

Speech production in XCR-1 is based on pre-recorded words that are stored in a non-volatile analog EEPROM chip, type ISD1000. (This type is now discontinued and replaced by the ISD1400 series devices.) This chip contains also a low-power audio amplifier that drives directly a small loudspeaker. The words are stored in this chip as temporal sequences of analog samples and consequently no analog-digital and digital-analog conversions are used. Nevertheless, the memory locations are addressable with eight address lines. In principle, the address is a binary number and each address count equals 0.125 seconds of stored sound. In XCR-1 four most significant address bits are used; this gives a time period of 2 seconds. This is more than enough for one word. The total capacity of the chip is 20 seconds, thus the maximum of ten words can be stored.

The use of shorter time period would be possible and this would allow a larger vocabulary. In this case, however, the auditory system has only four parallel neural signal lines, which equal four bits and this limits the number of possible words to fifteen. Thus it has been possible to drive the ISD1000 chip simply and directly with the four auditory neural signals, without any address decoding circuitry. This situation is equivalent to that where a speech producing organ is associatively controlled by the auditory neural signals.

It should be noted that the actual speech production method is of no theoretical significance here. Speech could also be produced by synthesis, but in the framework of this experiment it would only complicate the circuitry without providing any new insights. The important issue here is the grounding of meaning of the produced words. This is discussed in the subchapter 21.8., "Self-Talk".

21.4. The Visual Module

21.4.1. *Visual Perception/Response Feedback Loops*

The visual perception/response feedback loop module of the robot XCR-1 is designed for the purpose of searching and detecting active targets and their relative directions in respect to the robot heading.

The visual perception/response feedback loop module produces two kinds of information, namely the object feature information and the object direction information. The object feature information is broadcast to the auditory and emotional modules for naming and emotional significance evaluation and the direction information is broadcast to the wheel motor module so that the robot can find its way to the object. The visual perception/response feedback loop module receives information from the auditory module and the emotional module. The visual perception/response feedback loop module is depicted in Fig. 21.4.

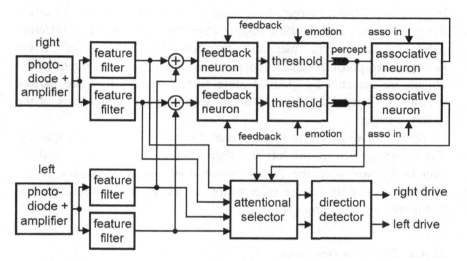

Fig. 21.4. The visual perception/response feedback loop module consists of two visual feature perception/response feedback loops, an attentional selector and direction detector. The direction detector provides information to the wheel motor drive circuits.

The visual perception/response feedback loop module utilizes two large area photodiodes (BPW34) as visual sensors. These are located in front of the robot and their distance is 30 mm from each other. Short focal length lenses (from cheap single use cameras) project the images of the targets on the photodiodes allowing improved detection and direction sensitivity. Each photodiode has its own amplifier.

The circuitry of visual perception/response feedback loop module consists of two visual feature perception/response feedback loops, an attentional selector and direction detector. The direction detector determines the direction of the attended object. The detailed operation of these circuits is described in the following.

21.4.2. *Object Recognition*

This robot is designed to detect visually active target objects. The active targets transmit pulsed infrared light and have different pulse frequencies. These frequencies constitute the detected visual features. Each visual feature is represented by its neural feature signal and each neural feature signal has its own perception/response feedback loop. Initially two different targets have been used and these targets are called the green object and the blue object. Thus there are only two different pulse frequencies in use and therefore the number of visual perception/response feedback loops is two. The pulse frequencies of the targets are detected by visual feature filters. These filters are narrow bandwidth filters that are tuned to the specific pulse frequencies of the different targets.

Both photodiodes have visual feature filters for each pulse frequency. The information from the left and right photodiode is summed and the sum is forwarded to the feature perception/response feedback loops, which process only feature information, as the direction information is lost in the summing process.

21.4.3. *Direction Detection*

Direction detection is based on the parallax effect, Fig. 21.5. If the pursued object is directly in front of the robot, then both photodiodes generate equal signals. If the object is on the left side, then due to the parallax effect the projected image of the target falls only partially on the right photodiode and the produced signal is weaker. At the same time, the projected image of the target covers more of the sensitive area of the left photodiode and the produced signal is stronger. In principle, these signals could be directly used via buffer amplifiers to drive the wheel motors, right signal would drive the left motor and vice versa. If the object is on the left side, then the left signal would be stronger and the right motor would run faster. This would turn the robot towards left and eventually the object would be directly in front of the robot. At that moment both photodiode signals would be equal and both wheel motors would run with the same speed. In this way a closed feedback control loop would arise whenever the robot is approaching a target. This feedback control loop would ensure that the robot would eventually reach a position, where the target would be quite exactly between the robot's gripper hands.

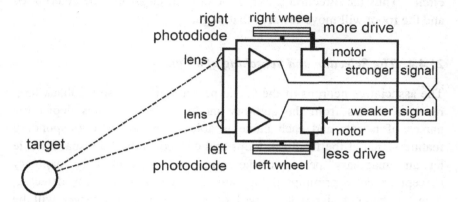

Fig. 21.5. Driving towards a target. Due to the parallax effect the projected image of the target falls only partially on the right photodiode and the produced signal is weaker. The stronger signal from the left photodiode drives the right wheel and the weaker signal from the right photodiode drives the left wheel. Consequently, the robot turns towards the target.

However, this direct method has a drawback. When the detected object is far away, the photodiode signals are very weak and the robot would move very slowly if at all. Therefore the distance information and the direction information must be separated from each other and only the direction information that does not depend on the distance, must be used to control the wheel motors. This is done by the direction detector circuit, which compares the relative strengths of the photodiode signals and based on these, produces full amplitude wheel motor drive signals regardless of the distance of the target. (Similar principle could be used for the sound direction detection.)

The robot may be seeing more than one target (in this case two) at the same time. Therefore an attentional selection mechanism is necessary. Artificial "saccades" allow the initial focusing of attention on each object in turn. The visual perception/response feedback loops are able to determine, which object is to be attended to and this is where the final decision is made. Attention in the visual module is controlled by emotional values and possible external verbal commands. The attended percept will control the selector, which will then pass only the corresponding feature filter output signals to the direction detector circuit. Thus the direction towards the desired target will be determined and the robot will move towards that target.

21.4.4. *The Selection and Searching of a Target*

The associative neurons of the visual perception/response feedback loop receives signals from the auditory module. These signals depict the names of targets and each name is able to evoke the corresponding feature signal at the feedback loops and consequently the visual module has an "imaginary" percept of the corresponding target. This imaginary percept executes priming; if the same target is now actually detected, then the percept signal level will be elevated and this target will be favored. The selected percept feature controls also the direction detection and consequently the direction towards the named target will be determined.

A target may be selected also by its emotional value. The emotional system controls the threshold levels at the outputs of the feedback neurons. The threshold level for a "good" feature is lowered and consequently the "good" target is favored over a neutral one.

21.5. The Emotional Module

21.5.1. *Pain and Pleasure*

The emotional system of the robot XCR-1 is based on the concepts of "pain" and "pleasure". "Pain" is seen as a harmful state that the system tries to interrupt and avoid. "Pleasure" is seen as a beneficial state that the system tries to pursue and sustain. Pain and pleasure are mediated by corresponding signals, but the phenomenal feel would be related to the actual system reactions. It is not claimed that XCR-1 would currently have any real phenomenal feel of pain or pleasure; the purpose is only to investigate pain-like and pleasure-like system behavior and emotional motivation in a hardwired associative architecture.

In XCR-1 pain reactions are initiated by shock vibrations in the robot frame. These vibrations are detected by a shock sensor that is bolted to the body of the robot. This sensor consists of a miniature magnetic earphone (from a toy), a low-pass filter amplifier and a threshold circuit. Due to the low-pass filtering and conditioning of the sensor signal, the shock sensor is not sensitive to normal external sounds or to the sounds that are generated by the motions of the robot itself. The pain-signal can only be activated by hitting the robot or with a very loud bang. The hardwired pain reactions include the release of any gripped object, backing off or reversing the motion direction. If the robot stands still and is hit from behind, the robot will move forward a little bit.

A "petting sensor" on top of the robot is used to initiate pleasure states. This sensor is sensitive to touch.

The signals from the "petting" and shock sensors are forwarded to "pleasure" and "pain" neurons. These neurons are used to associate positive emotional value, "pleasure", and negative emotional value, "pain", with visual percepts.

21.5.2. *Non-Events*

The emotional module contains a "non-event" detector, which senses extended periods of non-activity. This corresponds loosely and functionally to the feeling of boredom when nothing is happening. When a non-event state is detected, the search routine is initiated. If the robot has been gripping an object during the non-event state, the gripper will release and the robot will back off before initiating the forward-turn-forward-turn search routine.

21.5.3. *Emotional Significance*

Petting and hitting can be used to change the instantaneous behavior of the robot by the evoked emotional reactions. Better overall behavior would follow, if the applied petting and hitting, reward and punishment, could be remembered and associated with the corresponding activity. In this way the robot would learn, which actions were desirable and worth pursuing and which ones should be avoided; the robot would accumulate a track of positive and negative emotional significance.

In XCR-1 the function of emotional significance is facilitated by the pleasure and pain neurons in the style of the chapter 17. These neurons associate pain and pleasure with the objects that are related to pain and pleasure producing events. In this way these objects will be accompanied by emotional values, which will change the behavior of the robot by evoking the corresponding system reactions, when the object is encountered. The emotional values will also act as *motivational factors* before the objects are encountered; the robot will search objects with positive emotional value and avoid objects with negative emotional value.

21.5.4. *Self-Concept*

Sensory percepts can relate to the objects, their relationships and actions of the external world or to the perceiving subject itself, to its own body, own actions, own feelings and thoughts. The percepts of the subject itself allow the emergence of a self-concept, which is a necessary requirement for self-consciousness. This requirement has been recognized, for instance, by Damasio and his proposition is that at the basic level the sense of self is grounded to the representations (percepts) of the body [Damasio 2000, p. 22].

In the robot XCR-1, motion, touch and "pain" percepts are used to evoke a self-concept. It is the robot itself that is moving, it is the robot itself that feels the touch and it is the robot itself that is in pain; the presence of these percepts is always related to the experiencing self, which is the common denominator for these percepts. In XCR-1 the evoked self-concept is in the form of a self-concept signal. This signal allows the use of a verbal self-symbol, in this case the word "me". The robot may utter: "me hurt", when hit, "me search", when searching for an object and "me touch", when the touch sensors indicate contact pressure. These utterances may be followed by the name of the searched or touched object.

Current wiring and vocabulary does not allow the reports like "me see (something)" or "me hear (something)", but these would be simple additions along the already implemented lines.

21.6. The Gripper Module

The gripper mechanism allows the robot to grip suitably sized objects. The mechanism is depicted in Fig. 21.6.

Fig. 21.6. The gripper mechanism. The rotary movement of the motor spindle is transformed into the linear motion of a slider by the worm shaft. The gripper arms are operated by levers that are connected to the slider.

The gripper mechanism consists of two gripper arms that are driven by a worm shaft that is coupled directly to the gripper drive motor spindle via a rubber tubing joint. The rotary movement of the motor spindle is transformed into the linear motion of a slider by the worm shaft. The gripper arms are operated by levers that are connected to the slider.

The gripper hands are covered by pieces of conductive foam plastic, which function as touch sensors. The resistance of the conductive foam depends on the applied pressure and this can be easily sensed and translated into a corresponding voltage value by a high-impedance sense amplifier. In XCR-1 on/off touch sensing is used by applying a threshold circuit at the output of the sense amplifier. (Black conductive foam plastic sheets can sometimes be found in semiconductor packages. There are two varieties, brittle and soft. The soft variety is useful in this application.)

An optical sensor (gripper sensor) detects objects between the gripper hands. When an object is detected between the gripper hands, the robot will stop if it has been moving. Next, the gripper hands will close, unless this reaction is overridden by cognitive control. The gripper motor will stop when the touch pressure at both hands reaches a set limit. At that point the gripper holds the object quite firmly, but not too hard. There are limit switches at both ends of the worm shaft drive (these are micro switches from a dead PC mouse). These switches stop the gripper motor when the worm shaft drive reaches its extreme positions. The principle of the gripper circuit is shown in Fig. 21.7.

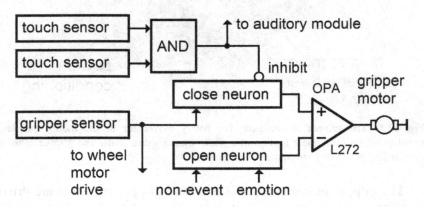

Fig. 21.7. The gripper circuit. The power operational amplifier (OPA) drives positive or negative voltage to the motor. Positive voltage closes the gripper and negative voltage opens it.

The gripper circuit consists of two neurons that are connected to a power operational amplifier (OPA). The excitation of the "close neuron" causes the OPA to drive positive voltage to the gripper motor and this will close the gripper arms. Touch sensor signals will inhibit the "close neuron" and the gripper motor will stop. The "open neuron" is an OR neuron and is controlled by the non-event detector and negative emotion. The excitation of the "open neuron" causes the OPA to drive negative voltage to the gripper motor and this will open the gripper arms.

21.7. The Wheel Drive Module

The robot XCR-1 has a two wheel drive system. The wheels are originally furniture wheels with ball bearings and rubber rim. Each wheel has its own DC-motor and the shaft of the motor drives directly the wheel's rubber rim. This is a simple and compact arrangement that requires no gearboxes and consequently the overall operation is very silent.

The wheel motors (type FF-180PH) are silent audio quality cassette load motors from surplus car C-cassette player mechanisms. These motors have good torque and only 50 mA no-load current, therefore they can be easily driven directly by one ampere-rated power operational amplifiers (such as L272). The wheel motor drive circuit for each wheel is given in Fig. 21.8.

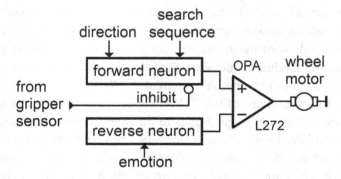

Fig. 21.8. The wheel motor drive circuit. The circuit is similar for both motors.

The wheel motors are controlled by forward and reverse neurons. These neurons cause the power operational amplifier to drive positive or negative voltage to the motor, which will then run forwards or backwards. The forward neurons are activated by the direction signals from the visual direction detector and the sequential search routine signals that are a response to a detected non-event state. The gripper sensor outputs a signal, when a target is located right between the gripper hands. At that moment any forward motion must be stopped to avoid collision and to allow the gripper hands to grip the target properly. This is achieved by inhibiting the output of the forward neuron by the gripper sensor signal. However, the gripper sensor signal does not inhibit a possible reverse motion.

Reverse motion is caused by the activation of the reverse neurons. These neurons receive signals from the emotional module.

21.8. Self-Talk

The self-talk in the XCR-1 robot is a running report of the instantaneous content of the neural system. The self-talk consists of a limited number of natural language (English) words that are produced by the ISD1000 analog audio memory chip in the previously described way by letting the signal patterns in the auditory module act as the binary addresses for the ISD1000 chip. The auditory signal patterns inside the auditory module are evoked associatively by broadcasts from the other modules and therefore act as symbols for the activity in the other modules. The SD1000 chip transforms the neural signal pattern symbols into natural language words with meanings that are grounded to the active percepts of the various modules. Therefore the self-talk offers a window into the inner workings of the cognitive system of the XCR-1.

The various modules of the XCR-1 robot broadcast continuously and simultaneously their percept signals to the other modules and also to the auditory module. However, only one word can be uttered at a time and therefore the broadcasts must be accepted serially, only one broadcast at a time. This function is realized by an attention threshold circuit at the associative input of the auditory module.

Several possibilities for this kind of attentional selection are available. The word order could be determined by inner models and emotional values or a fixed word order may be used. Flexible word orders would be desirable in more complicated systems. In this simple robot fixed word order works quite satisfactorily and is easy to implement with minimal hardware. Here the attention threshold circuit is made to "scan" the incoming broadcasts sequentially and accept only one broadcast at a time. This broadcast signal will then evoke the corresponding word. The implemented word order corresponds to the order of subject – verb – object – adjective.

New words cannot be initiated before the completion of the current word, therefore the attention threshold circuit can accept the next broadcast only after the completion of the previous word. This problem is solved by the end-of-word signal that is provided by the ISD1000 chip. This signal is made to control the scan timing. Fall-back timing is used when no words are uttered and consequently no end-of-word signals are available.

The self-talk begins as soon as the XCR-1 perceives something. The percepts of the "green" and "blue" test targets elicit word sequences like "me search green" or "me search blue", when the robot starts to move towards the detected target. ("Green" and "blue" are the names of the test targets. For the benefit of an observer, the targets have green and blue bands and in the robot the visual perception of "green" and "blue" targets are indicated by green and blue leds; the robot itself does not utilize these features.) When a target has been captured between the gripper hands, a word sequence like "me touch green" is produced. Hitting the robot will elicit the comment "me hurt". Emotional values can be associated with the test targets. These will we reported as "green bad" or "green good".

The simple self-talk of the XCR-1 robot reports the activity of all modalities in a symbolic way within the auditory modality. In doing so, the self-talk of XCR-1 illuminates the practical issues of basic grounding of word meaning. Complete natural language is more than that, but without this first step there cannot be any language. The next experimental steps would be the implementation of advanced sentences and the utilization of self-talk in planning and reasoning according to the multimodal model of language [see Haikonen 2003, 2007].

21.9. Cross-Associative Information Integration

21.9.1. *The Verbal-Corporal Teaching Experiment*

The XCR-1 robot recognizes some words that are associated with certain objects and these associations allow the verbal evocation of the "imagined" visual percepts of the corresponding objects without the actual presence and observation of these objects. For instance, the word "green" will evoke the virtual visual percept of the green object and the word "blue" will evoke the virtual visual percept of the blue object. Virtual percepts can be utilized in training instead of the actual encounter of real objects. For instance, it would be more safe to teach in advance, that a certain object would be dangerous and should be avoided as the actual presence of this object would already pose a risk. In the CXR-1 robot this can be done by the combined verbal-corporal teaching. Here the virtual percept of an object is evoked by its name and an emotional value is associated with this object by the simultaneous application of corporal "punishment" or "reward" (pain or pleasure) via the shock or petting sensors.

Later on, when the robot actually encounters and observes these objects visually, it will behave according to the associated emotional value of that object; it will avoid bad objects and favor good ones.

A simple experiment demonstrates this process. In this example the word "green" is used to evoke the corresponding "inner image" of the green object. At the same time corporal punishment is given; this will evoke the emotional value of <bad>, which will be associated with the "inner image" of the green object. The following partly overlapping events take place in this experiment.

1. The auditory perception of a name of an object; "green"
2. The evocation of an "imagined" visual percept of the object by the perceived name; <imagined green>
3. The perception of the simultaneous corporal reward or punishment; <punishing shock>
4. The evocation of the corresponding emotional value; <bad>

5. The association of the emotional value with the "imagined" percept of the object
6. A verbal self-report; "me hurt"
7. Fleeting inner "imagery" evoked by the self-report

This verbal teaching will alter the behavior of the robot. After the teaching episode the robot will no longer approach green objects in the normal way, instead it will be avoiding them. The events during a response are:

1. The visual perception of an object; <visual green>
2. The evocation of the emotional value of the percept; <bad>
3. The corresponding motor response change; <back off>
4. A verbal self-report; "green bad"
5. Fleeting inner "imagery" evoked by the self-report

Due to the principles of the perception/response feedback loop, the <bad> value that is associated with the internally evoked "inner image" of green, applies also to the visually perceived green.

21.10. Consciousness in the XCR-1

It is not claimed here that the robot XCR-1 were conscious. The main purposes of this project have been the verification of the feasibility of associative information processing and the main principles of the Haikonen Cognitive Architecture (HCA). However, XCR-1 does satisfy certain requirements for consciousness, at least to a minimal degree. In the previous chapters some basic requirements for conscious cognitive systems were identified. These requirements include:

1. The internal phenomenal appearance of neural activity, qualia
2. Direct sub-symbolic perception process that supports amodal features.
3. The externalization of non-contact percepts
4. Sensorimotor integration

5. Attention
6. Introspection of mental content in terms of sensory features, imagination, via feedback loops
7. Responses and reports (e.g. self-talk)
8. Retrospection and personal history
9. Emotional control and motivation
10. Somatosensory system for the grounding of the self-concept
11. The distinction between conscious and non-conscious operations
12. The transition from sub-symbolic to symbolic processing

The requirement #1 relates to the question about the possible internal appearance of the neural activity as qualia. It is obvious that there is neural activity inside the robot and it is also obvious that this neural activity does not appear to the robot as electric signals. But does it appear as percepts of external objects or heard sounds? The intuitive answer would seem to be *no*, but this is not certain; the sensory information is direct, it has effects on the system and the system is able to report it. The system is not able to perceive this information as electric signals; then, how does the system perceive it? At this moment this question remains as a philosophical one without a definite empirical answer. If a definite *yes* could be given, then the robot XCR-1 would be conscious, however with very limited situation awareness and with qualia that are definitely different from the human qualia. Nevertheless, the problem of consciousness would have been solved.

The requirement #2 follows from the basic requirement #1 as it states possible requirements for qualia. These requirements are fulfilled, at least partly, in the robot XCR-1. The perception process is direct and sub-symbolic.

The requirement #3 calls for the externalization of non-contact sensory percepts. This relates to the requirement #1; do the visual and auditory percepts appear internally in a way that could be externalized? The externalization effect itself would not seem to be very complicated, because an apparent location could be associated with a percept, which in itself does not inherently have one. In XCR-1 this effect might manifest itself in the detection of the direction of a target. The complete fulfillment of this requirement depends partly on the fulfillment of the first requirement.

The requirement #4, sensorimotor integration, is fulfilled in the robot XCR-1 within the limits of its construction. Visual perception is associatively and seamlessly connected with motor systems allowing the immediate utilization of visual direction information. Also pain and pleasure perception is connected with motor systems.

The requirement #5, attention, is realized in various forms in the robot XCR-1. Attention is effected by threshold controls. These, in turn, are controlled by emotional significance, virtual saccades and verbal expressions.

The requirement #6, introspection of mental content in terms of sensory features and imagination is fulfilled in the robot XCR-1, again within the limits of its construction. Introspection is facilitated by perception/response feedback loops in the way that is described earlier in this book. Imaginative visual percepts of objects that are not present can be evoked by spoken names.

The requirement #7, responses and reports (e.g. self-talk) is fulfilled in the robot XCR-1. There is a limited natural language self-talk that reports the inner "mental" content of the robot.

The requirement #8, retrospection and personal history is not currently implemented in XCR-1, because the limited number of neurons does not allow the accumulation of personal history. The robot can learn simple associations, but it is not able to make memories of its actions. This shortcoming could be easily fixed by the addition of further neuron groups.

The requirement #9, emotional control and motivation is implemented in XCR-1, again in a limited way. The learning of emotional values of objects can modify the behavior of the robot and these values can be used to motivate the robot to seek or avoid the objects.

The requirement #10, somatosensory system for the grounding of the self-concept is implemented in the robot XCR-1. The robot has the word "me" that is associated with various self-related percepts and situations, such as "me hurt", "me search", "me touch".

The requirement #11, the distinction between conscious and non-conscious operations, is inherently implemented in the robot XCR-1. Conscious operations would be those that are reported, while non-conscious operations would be those that take place without global attention and reporting. It should be noted that this kind of distinction may take place even if the system did not fulfill the requirement #1. Consequently, the sole fulfillment of this requirement does not guarantee any consciousness.

The requirement #12, the transition from sub-symbolic to symbolic processing is fulfilled in the robot XCR-1. The robot is able to understand some verbal words and also produce them in the form of neural signal patterns within the auditory perception/response feedback loop. These neural signal patterns in themselves are sub-symbolic representations for combinations auditory features, but here they are used as symbols for various entities from other sensory modalities. This realizes the transition from sub-symbolic to symbolic processing. For the benefit of an observer, these signal patterns are transformed into spoken words by the ISD1000 chip, but this does not add or remove anything from the general principle.

The XCR-1 is a simple robot and as a such its circuitry and system cannot support extended cognitive functions that were useful in the evaluation of possible preconditions for consciousness. However, one should not confuse the contents of consciousness and the mechanism of consciousness. The contents of human consciousness is rich and complicated, while the contents of consciousness in animals may be limited. Yet, the phenomenal mechanism behind consciousness is most probably the same and therefore also simple hardware system experiments might yield useful information about the phenomenal underpinnings of consciousness. Therefore, even though you may not be able to discuss the latest issues of philosophy with the robot XCR-1 any time soon, the robot XCR-1 should have certain merits on its own right as a non-program, no microprocessor, sub-symbolic/symbolic neural system hands-on exercise towards the understanding of some basic issues of conscious machines.

Chapter 22

Concluding Notes

22.1. Consciousness Explained

The one and only real mystery of consciousness is the existence of the inner appearance of the neural activity in the brain. The brain is a neural network, where the information is carried and manipulated by the interaction of neurons, synapses and glia. This process is a physical one, but it does not appear internally as such. We do not perceive the operation of the brain as it actually is and can be externally observed; the concert of electric signals and chemical messaging. In fact, we do not directly perceive our brain at all. Instead, we perceive the world around us *apparently* as it is. Likewise, we also perceive our body, its feelings, our mental thoughts and imaginations *apparently* as they are, with no apparent action of the neurons.

This is the one and only real problem of consciousness: Why and how do some neural processes in the brain have this internal appearance instead of the appearance of what they really are? And why would there be an internal appearance at all? All the other problems of consciousness are related to the technicalities of cognition and can, in principle, be solved by standard already known engineering methods.

The internal appearance cannot be an inherent property of some biological neural processes, because all neural processes do not have this appearance. Most of the neural activity of the brain is sub-conscious, without any internal appearance what so ever. Therefore, there must be some specific condition that generates the internal appearance.

The complete problem of consciousness can be condensed into three questions:

1. How does some neural activity appear internally as the subjective experience?
2. How does the subject become aware of its mental content?
3. How does the impression of I, the perceiving, reasoning and wanting self, arise?

The explanation of consciousness proposed in this book can be summarized as the answers to these questions as follows:

1) Consciousness is the presence of *internal appearances*. Instead of perceiving some neural activities as they are, we perceive them as the appearance of the world qualities *apparently* as they are. Consciousness is related to perception; the neural activity that has an internal appearance is about percepts of the external world and the body. These percepts manifest themselves in the form of *qualia*. Qualia are the qualities of percepts and therefore there are no percepts without qualia. Qualia are direct, not symbolic. They do not require interpretation or any evocation of meaning. Colors are colors and pain is pain, they are self-explanatory.

In perception it is the *effect* that counts, not the actual physical composition of the carrying neural activity, which is not observed as such. Neural machinery as such cannot have an internal appearance of neurons, because in the brain there are no sensors that could perceive the neurons and their signals.

A metaphor may illuminate this: Consider yourself floating in the ocean. Waves come and go and you are tossed up and down. An external observer would see the form of the waves, but your experience is the effect that these waves exert on you, the up-down motion. In a similar way, qualia are the effect that the perceptual neural signals have on you; with the aid of modern instruments an external observer may see the form of your neural activity, but instead of that, your inside experience will be that of the effect on your neural system. *This effect is the internal appearance of percepts, their qualia* and it is related to the qualities of sensed physical stimuli. We may know why red and green look different, but we may not know why red looks like red. However, the phenomenon of amodal features may allow a peek on the quality and feel of amodal qualia.

Thus, consciousness is based on perception; only *perception-related neural activity mirrors sensed entities* and can have corresponding *grounded* internal appearance. In this way, only percepts can constitute contents of consciousness.

Qualia are generated inside the brain, but normally this is not their perceived location. Sounds and visual objects seem to be out there and body sensations seem to have a location on the surface of the body or inside the body. Thus qualia are *externalized*; an outside location is *seamlessly* associated with them. This can be done, because neural signals do not carry intrinsic information about their origin. This externalization creates *the appearance of an inspectable outside world*.

There is an additional requirement for conscious perception; the *reportability*. Conscious states are reportable states. The subject can report these to itself (and others, but this is not necessary) in one way or another, not necessarily in a linguistic form. In order to enable reporting, qualia must be remembered for a while. A transient percept that does not generate any memory trace, does not generate any effects either and cannot be reported.

2) According to the aforesaid, only percepts can have internal appearance and thus constitute *the contents of consciousness*. Thoughts and imaginations are not directly related to sensory perception; they do not originate at sensors and in the brain there are no sensors for thoughts. Nevertheless, we are aware of our thoughts. According to the proposed explanation the thoughts and imaginations become consciously perceived, because they are transformed into *virtual percepts* with limited qualia via *feedback loops* that return the inner signals back to the perception process. In this way, *verbal thoughts* appear in the form of *virtually heard silent inner speech* and *imaginations* appear as *virtually seen visual forms*. Without the inner speech we would not hear our thoughts. In this way we become aware of our mental content.

3) The *concept of self* arises from the perception of the body and its sensations and the perception of the mental content. The percepts of the body and the mental content are different from the percepts of the environment; the environment changes when we move, but our body and

mental content follow us wherever we go. In that sense we are a moving vantage point. The body generates needs, these are our needs. Also pain is ours. The appearance of the concept of self is not a hard problem once the basic mechanism of consciousness is explained.

These three points suffice for the explanation of the phenomenon of consciousness; the gap is closed. Further explanations, which were also proposed here, would be related to the technicalities of the contents of consciousness, situation awareness and cognition. They are important issues as such, but secondary to the basic explanation of consciousness.

22.2. The Conclusion

The conclusions of this book lead to the basic technical requirements for conscious robots. These requirements are:

- Direct and transparent sub-symbolic perception process that preserves amodal qualia
- Associative information processing that facilitates the transition from sub-symbolic to symbolic processing
- Seamless information integration and sensorimotor integration
- A suitable feedback architecture that realizes the integration.

The Haikonen Cognitive Architecture (HCA) is an approach that is designed along these requirements. The HCA is a dynamic reactive sub-symbolic/symbolic system, which cannot be implemented with microprocessors, even though some of its functions may be simulated by computer programs. Therefore dedicated associative neuron group microchips should be used. Unfortunately these are not yet available.

The advancement of the consciousness research depends on practical experiments for which we now have sufficient technology. Now it is the time to complement the millennia of philosophical contemplations on consciousness with technological empiricism. The author hopes that this modest book would inspire talented students, researchers and engineers towards that direction.

Bibliography

Abbott, L.F. [1999] Lapique's introduction of the integrate-and-fire model neuron (1907), *Brain Research Bulletin* **50** (5/6), 303–304.

Aleksander, I. [2009] Essential Phenomenology for Conscious Machines: A Note on Franklin, Baars and Ramamurthy: "A Phenomenally Conscious Robot", *APA Newsletter on Philosophy and Computers* **8**(2)

Aleksander, I. and Dunmall, B. [2003] Axioms and Tests for the Presence of Minimal Consciousness in Agents, in O. Holland (ed.), *Machine Consciousness* (Imprint Academic), pp. 7–18.

Anderson, J. A. [1995] *An Introduction to Neural Networks* (MIT Press).

Arrabales, R. Ledezma, A. and Sanchis, A. [2009] "CERA-CRANIUM: A Test Bed for Machine Consciousness Research", in *Proc. International Workshop on Machine Consciousness 2009* (Hong Kong).

Arrabales R., Ledezma A. and Sanchis A. [2010] The Cognitive Development of Machine Consciousness Implementations, *IJMC* **2**(2), 213–225.

Baars, B. J. [1988] *A Cognitive Theory of Consciousness* (Cambridge University Press).

Baars, B. J. [1997] *In the Theater of Consciousness* (Oxford University Press).

Baars, B. J. and Franklin, S. [2009] Consciousness is computational: The LIDA model of Global Workspace Theory, *IJMC* **1**(1), 23–32.

Balduzzi D. and Tononi G. [2009] Qualia: The Geometry of Integrated Information, *PLoS Comput Biol* **5**(8): e1000462. doi:10.1371/journal.pcbi.1000462

Bergson, H. [1988] *Matter and Memory* (Zone Books).

Block, N. [1995] On a Confusion about a Function of Consciousness, *The Behavioral and Brain Sciences* **18**(2), 227–287.

Boltuc, P. [2009] The Philosophical Issue in Machine Consciousness, *IJMC* **1**(1), 155–176.

Botvinick, M. and Cohen J. [1998] Rubber hands `feel' touch that eyes see, *Nature* **391**, 756.

Browne, W. N. and Hussey, R. J. [2008] Emotional Cognitive Steps Towards Consciousness, *IJMC* **1**(2), 203–211.

Chalmers, D. J. [1995a] The Puzzle of Conscious Experience, *Scientific American*, **237**(6), 62–68.

Chalmers, D. J. [1995b] Facing Up to the Problem of Consciousness, *JCS* **2**(3), 200–219.

Chella, A. [2008] Perception Loop and Machine Consciousness, *APA Newsletter on Philosophy and Computers* **8**(1), 7–9.

Churchland, P. [1993] *Matter and Consciousness* (MIT Press).

Corkill, D. D. [1991] Blackboard Systems, *AI Expert* **6**(9), 40–47.

Crick, F. [1994] *The Astonishing Hypothesis* (Simon & Schuster), p. 3.

Crick, F. and Koch C. [1992] The problem of Consciousness, *Scientific American* **267**(3), 152–160.

Damasio, A. [2000] *The Feeling of What Happens* (Vintage)

Dennett, D. [1987] Cognitive wheels: The frame problem of AI, in C. Hookway (ed.), *Minds, machines and evolution* (Baen Books), pp. 41–42.

Dennett, D. [2001] Are we explaining consciousness yet? *Cognition* **79**, 221–237

Duch W, Oentaryo R. J. and Pasquier M. [2008] Cognitive architectures: where do we go from here? in P. Wang, B. Goertzel and S. Franklin (eds.), *Frontiers in Artificial Intelligence and Applications, Vol. 171* IOS Press, pp. 122–136.

Estes, W. K. [1994] *Classification and Cognition* (Oxford University Press).

Fodor, J. [1975] *The Language of Thought* (Crowell).

Gallup, G. Jr., [1970] Chimpanzees: Self-recognition, *Science* 167, 86–87

Gamez, D. [2008] *The development and analysis of conscious machines*, Ph.D. Thesis. University of Essex, Computing Department. p. 25.

Gibson, J.J. [1966] *The Senses Considered as Perceptual Systems* (Houghton Mifflin).

Gregory, R. L. [1998] *Eye and Brain* (Oxford University Press), p.168.

Haikonen, P. O. [1999] *An Artificial Cognitive Neural System Based on a Novel Neuron Structure and a Reentrant Modular Architecture with implications to Machine Consciousness*, Doctoral Thesis. Series B: Research Reports B4. Helsinki University of Technology, Applied Electronics Laboratory; 1999.

Haikonen, P. O. [1999b] Finnish patent no 103304

Haikonen, P. O. [2003] *The Cognitive Approach to Conscious Machines* (Imprint Academic).

Haikonen, P. O. [2005] "You Only Live Twice: Imagination in Conscious Machines", in *Proc. Symposium on Next Generation approaches to Machine Consciousness: Imagination, Development, Inter-subjectivity, and Embodiment (AISB05)* pp. 19–25.

Haikonen, P. O. [2007] *Robot Brains; circuits and systems for conscious machines* (John Wiley & Sons).

Haikonen, P. O. [2009] Qualia and Conscious Machines, *IJMC* **1**(2), 225–234

Haikonen, P. O. [2010] An Experimental Cognitive Robot, in A. V. Samsonovich, K. R. Johannsdottir, A. Chella and B. Goertzel (eds.), *Biologically Inspired Cognitive Architectures 2010* (IOS Press) pp. 52–57

Haikonen, P. O. [2011] XCR-1: An Experimental Cognitive Robot Based on an Associative Neural Architecture, *Cognitive Computation* **3**(2), 360–366

Harnad, S. [1992] The Turing Test Is Not a Trick: Turing Indistinguishability Is a Scientific Criterion, *SIGART Bulletin* **3**(4), 9–10.

Harnad, S. and Scherzer, P. [2007] "First, Scale Up to the Robotic Turing Test, Then Worry About Feeling", in *AI and Consciousness: Theoretical Foundations and Current Approaches*, AAAI Fall Symposium Technical Report FS-07-01. pp. 72–77.

Hayes-Roth, B. [1985] A Blackboard Architecture for Control, *Artificial Intelligence* **26**(3), 251–321.

Hebb, D. O. [1949] *The Organization of Behavior* (Wiley).

Hesslow, G. [2002] Conscious thought as simulation of behaviour and perception, *Trends in Cognitive Sciences* **6**(6), 242–247.

Hinton, G. E., McClelland, J. L. and Rumelhart, D. E. [1986] Distributed representations, In D. E. Rumelhart, and J. L. McClelland (eds.), *Parallel Distributed Processing: Explorations in the Microstructure of Cognition, Vol. 1: Foundations* (MIT Press), pp. 77–109.

Holland, O. and Goodman, R. [2003] Robots with Internal Models: A Route to Machine Consciousness? in O. Holland (ed.), *Machine Consciousness* (Imprint Academic), pp. 77–109.

Holland, O., Knight, R. and Newcombe, R. [2007] A Robot-Based Approach to Machine Consciousness, in A. Chella and R. Manzotti (eds.), *Artificial Consciousness*. (Imprint Academic), pp. 156–173.

Hume, D. A. [2000] *Treatise of Human Nature*, ed. D. F. Norton & M. J. Norton (Oxford University Press).

Jain, A. K., Mao J. and Mohiuddin K. M. [1996] Artificial Neural Networks: A Tutorial, *Computer* **29**(3), 31–44.

Jackson, F. [1982] Epiphenomenal Qualia, *Philosophical Quarterly*, 32, 127–36

Kanerva, P. [1988] *Sparse Distributed Memory* (MIT Press).

Kawamura, K. and Gordon, S. [2006] "From intelligent control to cognitive control", in *Proc. 11ᵗʰ International Symposium on Robotics and Applications* (ISORA), (Budapest, Hungary).

Kinouchi, Y. [2009] A Logical Model of Consciousness on an Autonomously Adaptive System, *IJMC* **1**(2), 235–242.

Koch, Ch. and Tononi G. [2011a] A Test for Consciousness, *Scientific American* **304**(6), 26–29.

Koch, Ch. and Tononi G. [2011b] Testing for Consciousness in Machines, *Scientific American Mind* **22**(4), 16–17.

LeDoux, J. [1996] *The Emotional Brain* (Simon & Schuster).

Lesser, E., Schaeps, T., Haikonen, P. and Jorgensen, C. [2008]. "Associative neural networks for machine consciousness: Improving existing AI technologies", in *Proc. of IEEEI 2008*, pp. 011–015.

Levine, J. [1983] Materialism and qualia: the explanatory gap, *Pacific Philosophical Quarterly* 64: 354–361.

Lewis, M., Sullivan, M. W., Stanger, C. and Weiss, M. [1989] Self-development and self-conscious emotions, *Child Development* **60** (1), 146–156.

MacKay, D. G. [1992] Constraints on Theories of Inner Speech, in D. Reisberg (ed.), *Auditory Imagery* (Psychology Press), pp. 131–148.

Malsburg, v.d., C. [1997] The Coherence Definition of Consciousness, in M. Ito, Y. Miyashita and E. T. Rolls (eds), *Cognition, Computation and Consciousness* (Oxford University Press), pp. 193–204.

Manzotti, R. and Tagliasco V. [2007] "An Externalist Process-Oriented Framework for Artificial Consciousness", in *Proc. AI and Consciousness: Theoretical Foundations and Current Approaches*. AAAI Fall Symposium, Technical report FS-07-01. AAAI Press, (Menlo Park California) pp. 96–101.

Marchetti, G. [2006] A presentation of Attentional Semantics, *Cognitive Processing* 7(3).

Maslin, K. [2001] *An Introduction to the Philosophy of Mind* (Polity Press), p. 87.

Massimi M., Ferrarelli F., Huber R., Esser SK., Singh H. and Tononi G. [2005] Breakdown of Cortical Effective Connectivity During Sleep, *Science* 309(5744) 2228–2232.

Mavridis, N. and Roy, D. [2006] "Grounded Situation Models for Robots: Where words and percepts meet", in *Proc. IROS'2006*, pp. 4690–4697.

Miyawaki, Y., Uchida, H., Yamashita, O., Sato, M., Morito, Y., Tanabe, H. C., Sadato, N. and Kamitani, Y., [2008] Visual Image Reconstruction from Human Brain Activity using a Combination of Multiscale Local Image Decoders, *Neuron* 60(5), 915–929.

Morin, A. and Everett, J. [1990] Inner speech as a mediator of self-awareness, self-consciousness, and self-knowledge: an hypothesis, *New Ideas in Psychology* 8(3), pp. 337–356.

Nairne, J. S. [1997] *The Adaptive Mind* (Brooks/Cole Publishing Company), p. 209.

Nagel, T. [1974] What Is it Like to Be a Bat?, *Philosophical Review* 83, 435–50.

Naselaris, T., Prenger, R. J., Kay, K. N., Oliver, M. and Gallant J. L. [2009] Bayesian Reconstruction of Natural Images from Human Brain Activity, *Neuron* 63(6), 902–915.

Nii, H. P. [1986] The Blackboard Model of Problem Solving, *AI Magazine* 7(2), 38–53.

Nishimoto, S., Vu, A. T., Naselaris, T., Benjamini, Y., Yu, B. and Gallant, J. L. [2011] Reconstructing Visual Experiences from Brain Activity Evoked by Natural Movies, *Current Biology*, 21(19), 1641–1646.

Pavlov, I. P. [1927/1960] *Conditional Reflexes* (Dover Publications).

Pearson, K. [1911/2007] *The Grammar of Science* (Cosimo Inc.).

Samsonovich, A. V. [2010] Toward a Unified Catalog of Implemented Cognitive Architectures, in A. V. Samsonovich, K. R. Johannsdottir, A. Chella and B. Goertzel (eds.), *Biologically Inspired Cognitive Architectures 2010* (IOS Press), pp. 195–244.

Sanz R., López, I. and Bermejo-Alonso, J. [2007] A Rationale and Vision for Machine Consciousness, in A. Chella and R. Manzotti (eds.), *Artificial Consciousness* (Imprint Academic), pp. 141–155.

Sanz, R., López, I., Rodríguez, M. and Hernández, C. [2007] Principles for Consciousness in Integrated Cognitive Control, *Neural Networks* **20**(9), 938–946.

Shanahan, M. P. and Baars, B. J. [2005] Applying Global Workspace Theory to the Frame Problem, *Cognition* **98**(2), 157–176.

Shanahan, M. [2010] *Embodiment and the Inner Life* (Oxford University Press).

Sloman A. [2010] An Alternative to Working on Machine Consciousness, *IJMC* **2**(1), 1–18.

Sommerhof, G. [2000] *Understanding Consciousness* (Sage Publications).

Steels, L. [2003] Language Re-Entrance and the "Inner Voice", in O. Holland (ed.), *Machine Consciousness* (Imprint Academic), pp. 173–185.

Suddendorf, T., Addis, D. R. and Corballis, M. C. [2009] Mental time travel and the shaping of the human mind, *Phil. Trans. R. Soc. B* 2009 364, 1317–1324. doi: 10.1098/rstb.2008.0301

Takeno, J., Inaba K. and Suzuki T. [2005] "Experiments and examination of mirror image cognition using a small robot", in *Proc. 6th IEEE International Symposium on Computational Intelligence in Robotics and Automation (CIRA 2005)*, pp. 493–498.

Tononi, G., Edelman, G. M. and Sporns, O. [1998] Complexity and coherency: integrating information in the brain, *Trends in Cognitive Sciences* **2**(12), 474–484.

Tononi, G. [2004] An information Integration Theory of Consciousness, *BMC Neuroscience* 2004, 5:42. doi:10.1186/1471-2202-5-42

Tononi, G. [2008] Consciousness as Integrated Information: A provisional Manifesto, *Biological Bulletin* **215**(3), 216–142.

Turing, A. M. [1950] Computing Machinery and Intelligence, *Mind* LIX no 2236 433–460.

Wand, M. and Schultz, T. [2009] "Towards Speaker-Adaptive Speech Recognition based on Surface Electromyography", in *Proc. International Conference on Bio-inspired Systems and Signal Processing (Biosignals 2009)*, (Porto, Portugal).

Wolpert, D. M., Ghahramani, Z. and Jordan, M. I. [1995] An Internal Model for Sensorimotor Integration, *Science*, New Series **269**(5232), 1880–1882.

Zwaan, R. A. and Radvansky, G. A. [1998] Situation Models in Language Comprehension and Memory, *Psychological Bulletin* **123**(2), 162–185.

Index